Image Processing Using Pulse-Coupled Neural Networks

T. Lindblad J.M. Kinser

Image Processing Using Pulse-Coupled Neural Networks

Second, Revised Edition

With 140 Figures

 Springer

Professor Dr. Thomas Lindblad
Royal Institute of Technology, KTH-Physics, AlbaNova
S-10691 Stockholm, Sweden
E-mail: Lindblad@particle.kth.se

Professor Dr. Jason M. Kinser
George Mason University
MSN 4E3, 10900 University Blvd., Manassas, VA 20110, USA, and
12230 Scones Hill Ct., Bristow VA, 20136, USA
E-mail: jkinser@gmu.edu

Library of Congress Control Number: 2005924953

ISBN-10 3-540-24218-X 2nd Edition, Springer Berlin Heidelberg New York
ISBN-13 978-3-540-24218-5 2nd Edition Springer Berlin Heidelberg New York
ISBN 3-540-76264-7 1st Edition, Springer Berlin Heidelberg New York

This work is subject to copyright. All rights are reserved, whether the whole or part of the material is concerned, specifically the rights of translation, reprinting, reuse of illustrations, recitation, broadcasting, reproduction on microfilm or in any other way, and storage in data banks. Duplication of this publication or parts thereof is permitted only under the provisions of the German Copyright Law of September 9, 1965, in its current version, and permission for use must always be obtained from Springer. Violations are liable to prosecution under the German Copyright Law.

Springer is a part of Springer Science+Business Media.

springeronline.com

© Springer-Verlag Berlin Heidelberg 1998, 2005
Printed in The Netherlands

The use of general descriptive names, registered names, trademarks, etc. in this publication does not imply, even in the absence of a specific statement, that such names are exempt from the relevant protective laws and regulations and therefore free for general use.

Typesetting and prodcution: PTP-Berlin, Protago-T$_E$X-Production GmbH, Berlin
Cover design: *design & production* GmbH, Heidelberg

Printed on acid-free paper   SPIN 10965221   57/3141/YU   5 4 3 2 1 0

Preface

It was stated in the preface to the first edition of this book that image processing by electronic means has been a very active field for decades. This is certainly still true and the goal has been, and still is, to have a machine perform the same image functions which humans do quite easily. In reaching this goal we have learnt about the human mechanisms and how to apply this knowledge to image processing problems. Although there is still a long way to go, we have learnt a lot during the last five or six years. This new information and some ideas based upon it has been added to the second edition of our book

The present edition includes the theory and application of two cortical models: the PCNN (pulse coupled neural network) and the ICM (intersecting cortical model). These models are based upon biological models of the visual cortex and it is prudent to review the algorithms that strongly influenced the development of the PCNN and ICM. The outline of the book is otherwise very much the same as in the first edition although several new application examples have been added.

In Chap. 7 a few of these applications will be reviewed including original ideas by co-workers and colleagues. Special thanks are due to Soonil D.D.V. Rughooputh, the dean of the Faculty of Science at the University of Mauritius Guisong, and Harry C.S. Rughooputh, the dean of the Faculty of Engineering at the University of Mauritius.

We should also like to acknowledge that Guisong Wang, a doctoral candidate in the School of Computational Sciences at GMU, made a significant contribution to Chap. 5.

We would also like to acknowledge the work of several diploma and Ph.D. students at KTH, in particular Jenny Atmer, Nils Zetterlund and Ulf Ekblad.

Stockholm and Manassas,   *Thomas Lindblad*
April 2005   *Jason M. Kinser*

Preface to the First Edition

Image processing by electronic means has been a very active field for decades. The goal has been, and still is, to have a machine perform the same image functions which humans do quite easily. This goal is still far from being reached. So we must learn more about the human mechanisms and how to apply this knowledge to image processing problems. Traditionally, the activities in the brain are assumed to take place through the aggregate action of billions of simple processing elements referred to as neurons and connected by complex systems of synapses. Within the concepts of artificial neural networks, the neurons are generally simple devices performing summing, thresholding, etc. However, we show now that the biological neurons are fairly complex and perform much more sophisticated calculations than their artificial counterparts. The neurons are also fairly specialised and it is thought that there are several hundred types in the brain and messages travel from one neuron to another as pulses.

Recently, scientists have begun to understand the visual cortex of small mammals. This understanding has led to the creation of new algorithms that are achieving new levels of sophistication in electronic image processing. With the advent of such biologically inspired approaches, in particular with respect to neural networks, we have taken another step towards the aforementioned goals.

In our presentation of the visual cortical models we will use the term Pulse-Coupled Neural Network (PCNN). The PCNN is a neural network algorithm that produces a series of binary pulse images when stimulated with a grey scale or colour image. This network is different from what we generally mean by artificial neural networks in the sense that it does not train.

The goad for image processing is to eventually reach a decision on the content of that image. These decisions are generally easier to accomplish by examining the pulse output of the PCNN rather than the original image. Thus the PCNN becomes a very useful pre-processing tool. There exists, however, an argument that the PCNN is more than a pre-processor. It is possible that the PCNN also has self-organising abilities which make it possible to use the PCNN as an associative memory. This is unusual for an algorithm that does not train.

Finally, it should be noted that the PCNN is quite feasible to implement in hardware. Traditional neural networks have had a large fan-in and fanout. In other words, each neuron was connected to several other neurons. In electronics a different "wire" is needed to make each connection and large networks are quite difficult to build. The PCNN, on the other hand, has only local connections and in most cases these are always positive. This is quite plausible for electronic implementation.

The PCNN is quite powerful and we are just in the beginning to explore the possibilities. This text will review the theory and then explore its known image processing applications: segmentation, edge extraction, texture extraction, object identification, object isolation, motion processing, foveation, noise suppression and image fusion. This text will also introduce arguments to its ability to process logical arguments and its use as a synergetic computer. Hardware realisation of the PCNN will also be presented.

This text is intended for the individual who is familiar with image processing terms and has a basic understanding of previous image processing techniques. It does not require the reader to have an extensive background in these areas. Furthermore, the PCNN is not extremely complicated mathematically so it does not require extensive mathematical skills. However, the text will use Fourier image processing techniques and a working understanding of this field will be helpful in some areas.

The PCNN is fundamentally unique from many of the standard techniques being used today. Many techniques have the same basic mathematical foundation and the PCNN deviates from this path. It is an exciting field that shows tremendous promise.

Contents

1 **Introduction and Theory** 1
  1.1 General Aspects 1
  1.2 The State of Traditional Image Processing 2
      1.2.1 Generalisation *versus* Discrimination 2
      1.2.2 "The World of Inner Products" 3
      1.2.3 The Mammalian Visual System 4
      1.2.4 Where Do We Go From Here? 4
  1.3 Visual Cortex Theory 5
      1.3.1 A Brief Overview of the Visual Cortex 5
      1.3.2 The Hodgkin–Huxley Model 6
      1.3.3 The Fitzhugh–Nagumo Model 7
      1.3.4 The Eckhorn Model 8
      1.3.5 The Rybak Model 9
      1.3.6 The Parodi Model 10
  1.4 Summary 10

2 **Theory of Digital Simulation** 11
  2.1 The Pulse-Coupled Neural Network 11
      2.1.1 The Original PCNN Model 11
      2.1.2 Time Signatures 16
      2.1.3 The Neural Connections 18
      2.1.4 Fast Linking 21
      2.1.5 Fast Smoothing 22
      2.1.6 Analogue Time Simulation 23
  2.2 The ICM – A Generalized Digital Model 24
      2.2.1 Minimum Requirements 25
      2.2.2 The ICM 26
      2.2.3 Interference 27
      2.2.4 Curvature Flow Models 31
      2.2.5 Centripetal Autowaves 32
  2.3 Summary 34

3 Automated Image Object Recognition ... 35
- 3.1 Important Image Features ... 35
- 3.2 Image Segmentation – A Red Blood Cell Example ... 41
- 3.3 Image Segmentation – A Mammography Example ... 42
- 3.4 Image Recognition – An Aircraft Example ... 43
- 3.5 Image Classification – Aurora Borealis Example ... 44
- 3.6 The Fractional Power Filter ... 46
- 3.7 Target Recognition – Binary Correlations ... 47
- 3.8 Image Factorisation ... 51
- 3.9 A Feedback Pulse Image Generator ... 52
- 3.10 Object Isolation ... 55
- 3.11 Dynamic Object Isolation ... 58
- 3.12 Shadowed Objects ... 60
- 3.13 Consideration of Noisy Images ... 62
- 3.14 Summary ... 67

4 Image Fusion ... 69
- 4.1 The Multi-spectral Model ... 69
- 4.2 Pulse-Coupled Image Fusion Design ... 71
- 4.3 A Colour Image Example ... 73
- 4.4 Example of Fusing Wavelet Filtered Images ... 75
- 4.5 Detection of Multi-spectral Targets ... 75
- 4.6 Example of Fusing Wavelet Filtered Images ... 80
- 4.7 Summary ... 81

5 Image Texture Processing ... 83
- 5.1 Pulse Spectra ... 83
- 5.2 Statistical Separation of the Spectra ... 87
- 5.3 Recognition Using Statistical Methods ... 88
- 5.4 Recognition of the Pulse Spectra via an Associative Memory ... 89
- 5.5 Summary ... 92

6 Image Signatures ... 93
- 6.1 Image Signature Theory ... 93
  - 6.1.1 The PCNN and Image Signatures ... 94
  - 6.1.2 Colour Versus Shape ... 95
- 6.2 The Signatures of Objects ... 95
- 6.3 The Signatures of Real Images ... 97
- 6.4 Image Signature Database ... 99
- 6.5 Computing the Optimal Viewing Angle ... 100
- 6.6 Motion Estimation ... 103
- 6.7 Summary ... 106

7 Miscellaneous Applications 107
- 7.1 Foveation 107
  - 7.1.1 The Foveation Algorithm 108
  - 7.1.2 Target Recognition by a PCNN Based Foveation Model 110
- 7.2 Histogram Driven Alterations 113
- 7.3 Maze Solutions 115
- 7.4 Barcode Applications 116
  - 7.4.1 Barcode Generation from Data Sequence and Images 117
  - 7.4.2 PCNN Counter 121
  - 7.4.3 Chemical Indexing 121
  - 7.4.4 Identification and Classification of Galaxies 126
  - 7.4.5 Navigational Systems 131
  - 7.4.6 Hand Gesture Recognition 134
  - 7.4.7 Road Surface Inspection 137
- 7.5 Summary 141

8 Hardware Implementations 143
- 8.1 Theory of Hardware Implementation 143
- 8.2 Implementation on a CNAPs Processor 144
- 8.3 Implementation in VLSI 146
- 8.4 Implementation in FPGA 146
- 8.5 An Optical Implementation 151
- 8.6 Summary 153

References 155

Index 163

1 Introduction and Theory

1.1 General Aspects

Humans have an outstanding ability to recognise, classify and discriminate objects with extreme ease. For example, if a person was in a large classroom and was asked to find the light switch it would not take more than a second or two. Even if the light switch was located in a different place than the human expected or it was shaped differently than the human expected it would not be difficult to find the switch. Humans also don't need to see hundreds of exemplars in order to identify similar objects. For example, a human needs to see only a few dogs and then he is able to recognise dogs even from species that he has not seen before. This recognition ability also holds true for animals, to a greater or lesser extent. A spider has no problem recognising a fly. Even a baby spider can do that. At this level we are talking about a few hundred to a thousand processing elements or neurons. Nevertheless the biological systems seem to do their job very well.

Computers, on the other hand, have a very difficult time with these tasks. Machines need a large amount of memory and significant speed to even come close to the processing time of a human. Furthermore, the software for such simple general tasks does not exist. There are special problems where the machine can perform specific functions well, but the machines do not perform general image processing and recognition tasks.

In the early days of electronic image processing, many thought that a single algorithm could be found to perform recognition. The most popular of these is Fourier processing. It, as well as many of its successors, has fallen short of emulating human vision. It has become obvious that the human uses many elegantly structured processes to achieve its image processing goals, and we are beginning to understand only a few of these.

One of the processes occurs in the visual cortex, which is the part of the brain that receives information from the eye. At this point in the system the eye has already processed and significantly changed the image. The visual cortex converts the resultant eye image into a stream of pulses. A synthetic model of this portion of the brain for small mammals has been developed and successfully applied to many image processing applications.

So then many questions are raised. How does it work? What does it do? How can it be applied? Does it gain us any advantage over current systems?

Can we implement it with today's hardware knowledge? This is what many scientists are working with today [2].

1.2 The State of Traditional Image Processing

Image processing has been a science for decades. Early excitement was created with the invention of the laser, which opened the door for optical Fourier image processing. Excitement was heightened further as the electronic computer became powerful enough and cheap enough to process images of significant dimension. Even though many scientists are working in this field, progress towards achieving recognition capabilities similar to humans has been very slow in coming.

Emulation of the visual cortex takes new steps forward for a couple of reasons. First, it directly emulates a portion of the brain, which we believe to be the most efficient image processor available. Second, is that mathematically it is fundamentally different than many such traditional algorithms being used today.

1.2.1 Generalisation *versus* Discrimination

There are many terms used in image processing which need to be clarified immediately. Image processing is a general term that covers many areas. Image processing includes morphology (changing the image into another image), filtering (removing or extracting portions of the image), recognition, and classification.

Filtering an image concerns the extraction of a certain portion of the image. These techniques may be used to find all of the edges, or find a particular object within the image, or to locate particular object. There are many ways of filtering an image of which a few will be discussed.

Recognition is concerned with the identification of a particular target within the image. Traditionally, a target is an object such as a dog, but targets can also be signal signatures such as a certain set of frequencies or a pattern. The example of recognising dogs is applicable here. Once a human has seen a few dogs he can then recognise most dogs.

Classification is slightly different that recognition. Classification also requires that a label be applied to the portion of the input. It is possible to recognise that a target exists but not be able to attach a specific label to it.

It should also be noted that there are two types of recognition and classification. These types are generalisation and discrimination. Generalisation is finding the similarities amongst the classes. For example, we can see an animal with four legs, a tail, fur, and the shape and style similar to those of the dogs we have seen, and can therefore recognise the animal as a dog. Discrimination requires knowledge of the differences. For example, this dog may have a short snout and a curly tail, which is quite different than most other dogs, and we therefore classify this dog as a pug.

1.2.2 "The World of Inner Products"

There are many methods that are used today in image processing. Some of the more popular techniques are frequency-based filters, neural networks, and wavelets. The fundamental computational engine in each of these is the inner product. For example, a Fourier filter produces the same result as a set of inner products for each of the possible positions that the target filter can be overlaid on the input image.

A neural network may consist of many neurons in several layers. However, the computation for each neuron is an inner product of the weights with the data. After the inner product computation the result is passed through a non-linear operation. Wavelets are a set of filters, which have unique properties when the results are considered collectively. Again the computation can be traced back to the inner product.

The inner product is a first order operation which is limited in the services it can provide. That is why algorithms such as filters and networks must use many inner products to provide meaningful results for higher order problems. The difficulty in solving a higher order problem with a set of inner products is that the number of inner products necessary is neither known nor easy to determine, and the role of each inner product is not easily identified. Some work towards solving these problems for binary systems have been proposed [8]. However, for the general case of analogue data the user must resort to using training algorithms (many of which require the user to predetermine the number of inner products and their relationship to each other). This training optimises the inner products towards a correct solution. This training may be very involved, tedious, computationally costly and provides no guarantee of a solution.

Most importantly is that the inner product is extremely limited in what it can do. This is a first order computation and can only extract one order of information from a data set. One well known problem is the XOR (exclusive OR) gate, which contains four, 2D inputs paired with 1D outputs, namely (00:0, 01:1, 10:1, 11:0). This system can not be mapped fully by a single inner product since it is a second order problem. Feedforward artificial neural networks, for example, require two layers of neurons to solve the XOR task.

Although inner products are extremely limited in what they can do, most of the image recognition engines rely heavily upon them. The mammalian system, however, uses a higher order system that is considerably more complicated and powerful.

1.2.3 The Mammalian Visual System

The mammalian visual system is considerably more elaborate than simply processing an input image with a set of inner products. Many operations are performed before decisions are reached as to the content of the image. Furthermore, neuro-science is not at all close to understanding all of the operations. This section will mention a few of the important operations to provide a glimpse of the complexity of the processes. It soon becomes clear that the mammalian system is far more complicated than the usual computer algorithms used in image recognition. It is almost silly to assume that such simple operations can match the performance of the biological system.

Of course, image input is performed through the eyes. Receptors within the retina at the back of the eye are not evenly distributed nor are they all sensitive to the same optical information. Some receptors are more sensitive to motion, colour, or intensity. Furthermore, the receptors are interconnected. When one receptor receives optical information it alters the behaviour of other surrounding receptors. A mathematical operation is thus performed on the image before it even leaves the eye.

The eye also receives feedback information. We humans do not stare at images, we foveate. Our centre of attention moves about portions of the image as we gather clues as to the content. Furthermore, feedback information also alters the output of the receptors.

After the image information leaves the eye it is received by the visual cortex. Here the information is further analysed by the brain. The investigations of the visual cortex of the cat [1] and the guinea pig [12] have been the foundation of the digital models used in this text. Although these models are a big step in emulating the mammalian visual system, they are still very simplified models of a very complicated system. Intensive research continues to understand fully the processing. However, much can still be implemented or applied already today.

1.2.4 Where Do We Go From Here?

The main point of this chapter is that current computer algorithms fail miserably in attempting to perform image recognition at the level of a human. The reason is obvious. The computer algorithms are incredibly simple compared to what we know of the biological systems. In order to advance the computer systems it is necessary to begin to emulate some of the biological systems.

One important step in this process is to emulate the processes of the visual cortex. These processes are becoming understood although there still exists significant debate on them. These processes are very powerful and can instantly lead to new tools to the image recognition field.

1.3 Visual Cortex Theory

In this text we will explore the theory and application of two cortical models: the PCNN (pulse coupled neural network) and the ICM (intersecting cortical model) [3, 4]. However, these models are based upon biological models of the visual cortex. Thus, it is prudent to review the algorithms that strongly influenced the development of the PCNN and ICM.

1.3.1 A Brief Overview of the Visual Cortex

While there are discussions as to the actual cortex mechanisms, the products of these discussions are quite useful and applicable to many fields. In other words, the algorithms being presented as cortical models are quite useful regardless of their accuracy in modelling the cortex. Following this brief introduction to the primate cortical system, the rest of this book will be concerned with applying cortical models and not with the actual mechanisms of the visual cortex.

In spite of its enormous complexity, two basic hierarchical pathways can model the visual cortex system: the pavocellular one and the mangnocellular one, processing (mainly) colour information and form/motion, respectively. Figure 1.1 shows a model of these two pathways. The retina has luminance and colour detectors which interpret images and pre-process them before conveying the information to visual cortex. The Lateral Geniculate Nucleus, LGN, separates the image into components that include luminance, contrast, frequency, etc. before information is sent to the visual cortex (labelled V, in Fig. 1.1).

The cortical visual areas are labelled V1 to V5 in Fig. 1.1. V1 represents the striate visual cortex and is believed to contain the most detailed and least processed image. Area V2 contains a visual map that is less detailed and pre-processed than area V1. Areas V3 to V5 can be viewed as speciality areas and process only selective information such as, colour/form, static form and motion, respectively.

Information between the areas flows in both directions, although only the feedforward signals are shown in Fig. 1.1. The processing area spanned by each neuron increases as you move to the right in Fig. 1.1, i.e. a single neuron in V3 processes a larger part of the input image than a single neuron in V1.

The re-entrant connections from the visual areas are not restricted to the areas that supply its input. It is suggested that this may resolve conflict between areas that have the same input but different capabilities.

Much is to be learnt from how the visual cortex processes information, adapts to both the actual and feedback information for intelligent processing. However, a 'smart sensor' will probably never look like the visual cortex system, but only use a few of its basic features.

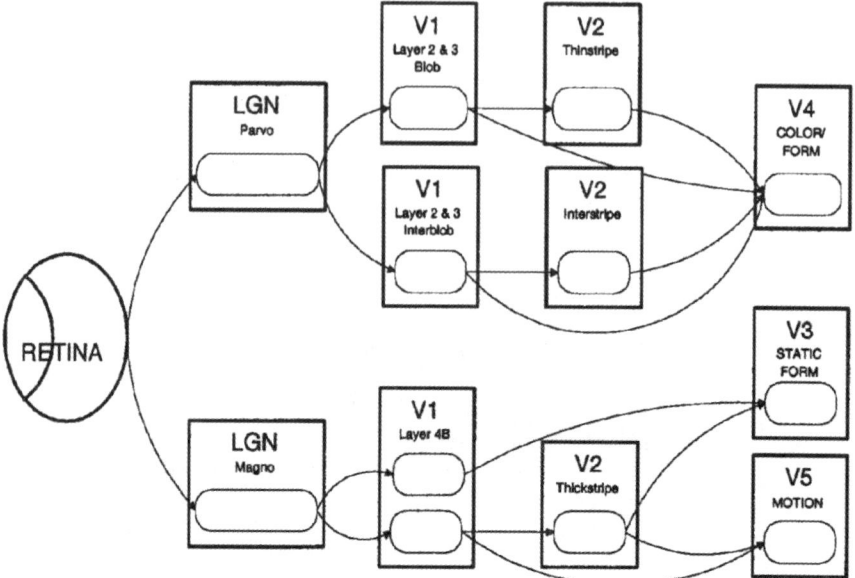

Fig. 1.1. A model of the visual system. The abbreviations are explained in the text. Only feedforward signals are shown

1.3.2 The Hodgkin–Huxley Model

Research into mammalian cortical models received its first major thrust about a half century ago with the work of Hodgkin and Huxley [6]. Their system described membrane potentials as

$$I = m^3 h G_{\text{Na}}(E - E_{\text{Na}}) + n^4 G_{\text{K}}(E - E_{\text{K}}) + G_{\text{L}}(E - E_{\text{L}}), \quad (1.1)$$

where I is the ionic current across the membrane, m is the probability that an open channel has been produced, G is conductance (for sodium, potassium, and leakage), E is the total potential and a subscripted E is the potential for the different constituents. The probability term was described by,

$$\frac{\mathrm{d}m}{\mathrm{d}t} = a_m(1 - m) - b_m m, \quad (1.2)$$

where a_m is the rate for a particle not opening a gate and b_m is the rate for activating a gate. Both a_m and b_m are dependent upon E and have different forms for sodium and potassium.

The importance to cortical modelling is that the neurons are now described as a differential equation. The current is dependent upon the rate changes of the different chemical elements. The dynamics of a neuron are now described as an oscillatory process.

1.3.3 The Fitzhugh–Nagumo Model

A mathematical advance published a few years later has become known as the Fitzhugh–Nagumo model [5,10] in which the neuron's behaviour is described as a van der Pol oscillator. This model is described in many forms but each form is essentially the same as it describes a coupled oscillator for each neuron. One example [9] describes the interaction of an excitation x and a recovery y,

$$\varepsilon \frac{\mathrm{d}x}{\mathrm{d}t} = -y - g(x) + I\,, \tag{1.3}$$

and

$$\frac{\mathrm{d}y}{\mathrm{d}t} = x - by\,, \tag{1.4}$$

where $g(x) = x(x-a)(x-1)$, $0 < a < 1$, I is the input current, and $\varepsilon \ll 1$. This coupled oscillator model will be the foundation of the many models that would follow.

These equations describe a simple coupled system and very simple simulations can present different characteristics of the system. By using ($\varepsilon = 0.3$, $a = 0.3$, $b = 0.3$, and $I = 1$) it is possible to get an oscillatory behaviour as shown in Fig. 1.2. By changing a parameter such as b it is possible to generate different types of behaviour such as steady state (Fig. 1.3 with $b = 0.6$).

The importance of the Fitzhugh–Nagumo system is that it describes the neurons in a manner that will be repeated in many different biological models. Each neuron is two coupled oscillators that are connected to other neurons.

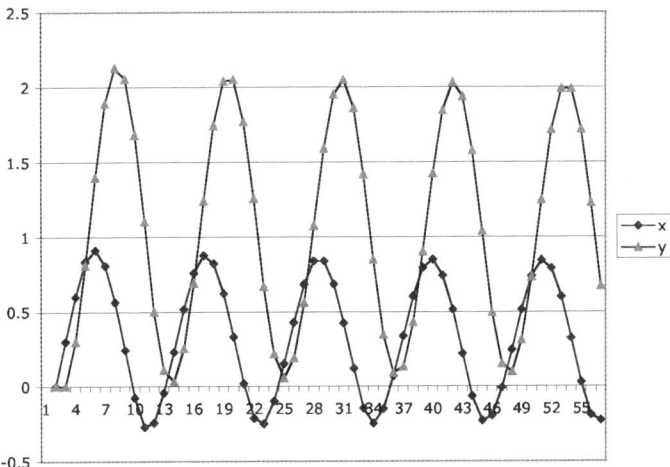

Fig. 1.2. An oscillatory system described through the Fitzhugh–Nagumo equations

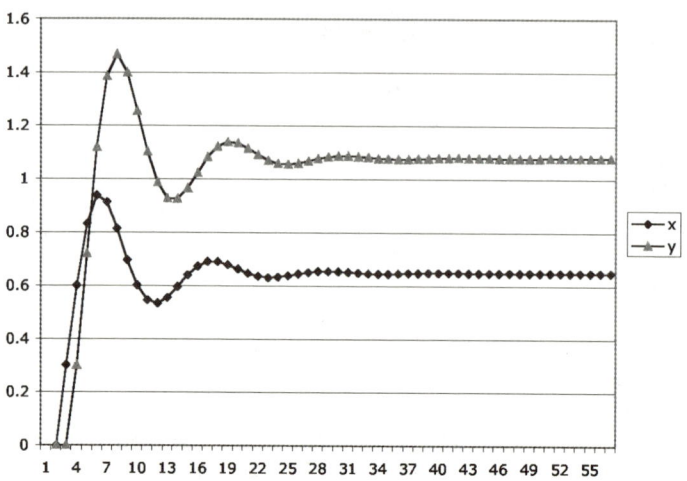

Fig. 1.3. A steady state system described through the Fitzhugh–Nagumo equations

1.3.4 The Eckhorn Model

Eckhorn [1] introduced a model of the cat visual cortex, and this is shown schematically in Fig. 1.4, and inter-neuron communication is shown in Fig. 1.5. The neuron contains two input compartments: the feeding and the linking. The feeding receives an external stimulus as well as local stimulus. The linking receives local stimulus. The feeding and the linking are combined in a second-order fashion to create the membrane voltage, U_m that is then compared to a local threshold, Θ.

The Eckhorn model is expressed by the following equations,

$$U_{m,k}(t) = F_k(t)[1 + L_k(t)] \tag{1.5}$$

$$F_k(t) = \sum_{i=1}^{N} \left[w_{ki}^f Y_i(t) + S_k(t) + N_k(t) \right] \otimes I\left(V^a, \tau^a, t\right) \tag{1.6}$$

$$L_k(t) = \sum_{i=1}^{N} \left[w_{ki}^l Y_i(t) + N_k(t) \right] \otimes I\left(V^l, \tau^l, t\right) \tag{1.7}$$

$$Y_k(t) = \begin{cases} 1 & \text{if } U_{m,k}(t) \geq \Theta_k(t) \\ 0 & \text{Otherwise} \end{cases} \tag{1.8}$$

where, in general

$$X(t) = Z(t) \otimes I(v, \tau, t) \tag{1.9}$$

is

$$X[n] = X[n-1]e^{-t/\tau} + V Z[n] \tag{1.10}$$

1.3 Visual Cortex Theory

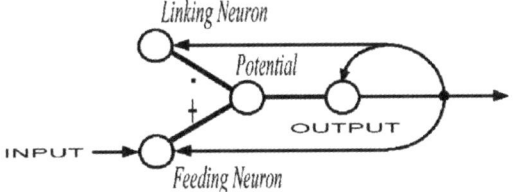

Fig. 1.4. The Eckhorn-type neuron

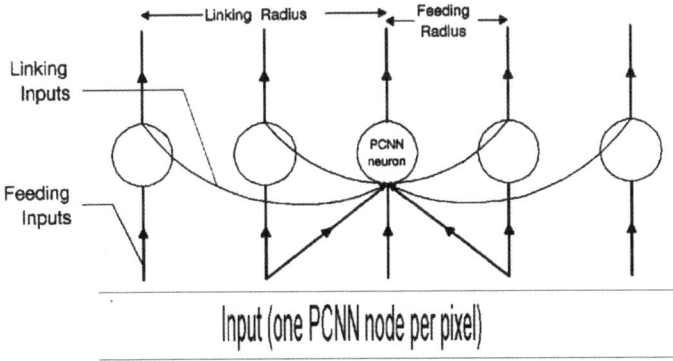

Fig. 1.5. Each PCNN neuron receives inputs from its own stimulus and also from neighbouring sources (feeding radius). In addition, linking data, i.e. outputs of other PCNN neurons, is added to the input

Here N is the number of neurons, w is the synaptic weights, Y is the binary outputs, and S is the external stimulus. Typical value ranges are $\tau^a = [10, 15]$, $\tau^l = [0.1, 1.0]$, $\tau^s = [5, 7]$, $V^a = 0.5$, $V^l = [5, 30]$, $V^s = [50, 70]$, $\Theta_o = [0.5, 1.8]$.

1.3.5 The Rybak Model

Independently, Rybak [12] studied the visual cortex of the guinea pig and found similar neural interactions. While Rybak's equations differ from Eckhorn's the behaviour of the neurons is quite similar. Rybak's neuron has two compartments X and Z. These interact with the stimulus, S, as,

$$X^S_{ij} = F^S \otimes \|S_{ij}\|, \tag{1.11}$$

$$X^I_{ij} = F^I \otimes \|Z_{ij}\|, \tag{1.12}$$

$$Z_{ij} = f\left\{\sum X^S_{ij} - \left(\frac{1}{\tau p + 1}\right) X^I_{ij} - h\right\}. \tag{1.13}$$

where F^S are local On-Centre/Off-Surround connections, F^I are local directional connections, τ is the time constant and h is a global inhibitor. In the cortex there are several such networks which work on the input at differing resolutions and differing F^I. The nonlinear threshold function is denoted $f\{\}$.

1.3.6 The Parodi Model

There is still great disagreement as to the exact model of the visual cortex. Recently, Parodi [11] presented alternatives to the Eckhorn model. The arguments against the Eckhorn model included the lack of synchronisation of neural firings, the undesired similar outputs for both moving and stationary targets and that neural modulations in the linking fields were measured considerably higher than the Eckhorn model allowed.

Parodi presented an alternative model, which included delays along the synaptic connections and would require that the neurons be occasionally reset *en masse*. Parodi's system followed these equations,

$$\frac{\partial V(x,y,t)}{\partial t} = -\frac{V(x,y,t)}{\tau} + D\nabla^2 V(x,y,t) + h(x,y,t), \qquad (1.14)$$

where V_i is the potential for the ith neuron, D is the diffusion ($D = a^2/C\ R_c$), R_c is the neural coupling resistance, $t = C\ R_l$, R_l is the leakage resistance, and $R_c^{-1} < R_l^{-1}$,

$$h_i(t) = \sum_j w_{ij} \delta(t - t_j^s - \tau_{ij}). \qquad (1.15)$$

1.4 Summary

Biological models of the visual cortex portray each neuron as a coupled oscillator with connections to other neurons. This differs significantly from traditional digital image processing approaches which tend to rely on first order mathematics. Building powerful image processing engines will require the use of more powerful engines and thus a cortical model will be employed for a variety of image processing applications in the subsequent chapters.

2 Theory of Digital Simulation

In this section two digital models will be presented. The first is the Pulse-Coupled Neural Network (PCNN) which for many years was the standard for many image processing applications. The PCNN is based solely on the Eckhorn model but there are many other cortical models that exist. These models all have a common mathematical foundation, but beyond the common foundation each also had unique terms. Since the goal here is to build image processing routines and not to exactly simulate the biological system a new model was constructed. This model contained the common foundation without the extra terms and is therefore viewed as the intersection of the several cortical models, and it is named the Intersecting Cortical Model (ICM).

2.1 The Pulse-Coupled Neural Network

The Pulse-Coupled Neural Network is to a very large extent based on the Eckhorn model except for a few minor modifications required by digitisation. The early experiments demonstrated that the PCNN could process images such output was invariant to images that were shifted, rotated, scaled, and skewed. Subsequent investigations determined the basis of the working mechanisms of the PCNN and led to its eventual usefulness as an image-processing engine.

2.1.1 The Original PCNN Model

A PCNN neuron shown in Fig. 2.1 contains two main compartments: the Feeding and Linking compartments. Each of these communicates with neighbouring neurons through the synaptic weights M and W respectively. Each retains its previous state but with a decay factor. Only the Feeding compartment receives the input stimulus, S. The values of these two compartments are determined by,

$$F_{ij}[n] = e^{\alpha_F \delta n} F_{ij}[n-1] + S_{ij} + V_F \sum_{kl} M_{ijkl} Y_{kl}[n-1], \qquad (2.1)$$

$$L_{ij}[n] = e^{\alpha_L \delta n} L_{ij}[n-1] + V_L \sum_{kl} W_{ijkl} Y_{kl}[n-1], \qquad (2.2)$$

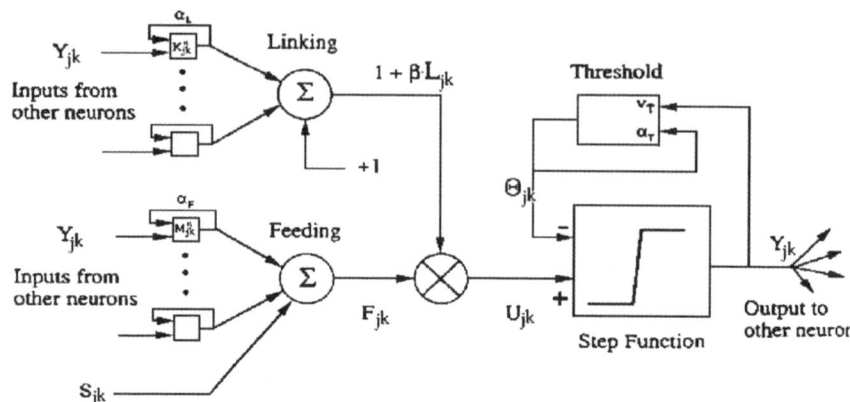

Fig. 2.1. Schematic representation of a PCNN processing element

where F_{ij} is the Feeding compartment of the (i,j) neuron embedded in a 2D array of neurons, and L_{ij} is the corresponding Linking compartment. Y_{kl}'s are the outputs of neurons from a previous iteration $[n-1]$. Both compartments have a memory of the previous state, which decays in time by the exponent term. The constants V_F and V_L are normalising constants. If the receptive fields of M and W change then these constants are used to scale the resultant correlation to prevent saturation.

The state of these two compartments are combined in a second order fashion to create the internal state of the neuron, U. The combination is controlled by the linking strength, β. The internal activity is calculated by,

$$U_{ij}[n] = F_{ij}[n]\{1 + \beta L_{ij}[n]\}. \tag{2.3}$$

The internal state of the neuron is compared to a dynamic threshold, Θ, to produce the output, Y, by

$$Y_{ij}[n] = \begin{cases} 1 & \text{if } U_{ij}[n] > \Theta_{ij}[n] \\ 0 & \text{Otherwise} \end{cases}. \tag{2.4}$$

The threshold is dynamic in that when the neuron fires $(Y > \Theta)$ the threshold then significantly increases its value. This value then decays until the neuron fires again. This process is described by,

$$\Theta_{ij}[n] = e^{\alpha_\Theta \delta n}\Theta_{ij}[n-1] + V_\Theta Y_{ij}[n], \tag{2.5}$$

where V_Θ is a large constant that is generally more than an order of magnitude greater than the average value of U.

The PCNN consists of an array (usually rectangular) of these neurons. Communications, M and W are traditionally local and Gaussian, but this is not a strict requirement. Initially, values of arrays, F, L, U, and Y are all set to zero. The values of the Θ elements are initially 0 or some larger value depending upon the user's needs. This option will be discussed at the

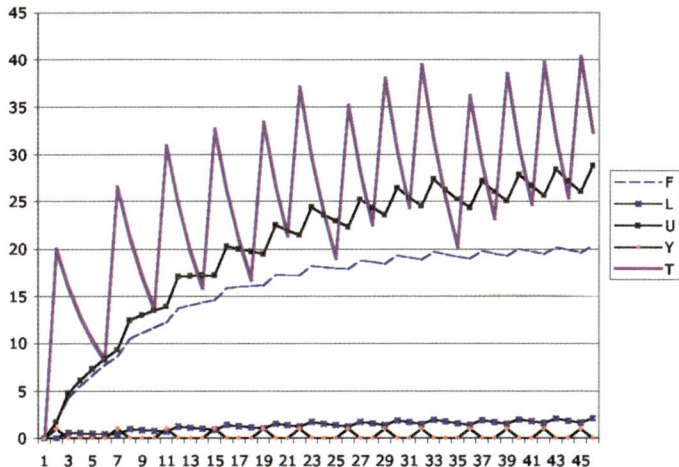

Fig. 2.2. An example of the progression of the states of a single neuron. See the text for explanation of L, U, T and F

end of this chapter. Each neuron that has any stimulus will fire in the initial iteration, which, in turn, will create a large threshold value. It will then take several iterations before the threshold values decay enough to allow the neuron to fire again. The latter case tends to circumvent these initial iterations which contain little information.

The algorithm consists of iteratively computing (2.1) through (2.5) until the user decides to stop. There is currently no automated stop mechanism built into the PCNN.

Consider the activity of a single neuron. It is receiving some input stimulus, S, and stimulus from neighbours in both the Feeding and Linking compartments. The internal activity rises until it becomes larger than the threshold value. Then the neuron fires and the threshold sharply increases then begins its decay until once again the internal activity becomes larger than the threshold. This process gives rise to the pulsing nature of the PCNN. Figure 2.2 displays the states within a single neuron embedded in a 2D array as it progresses in time.

In this typical example, the F, L, and U maintain values within individual ranges. The threshold can be seen to reflect the pulsing nature of the neuron.

The pulses also trigger communications to neighbouring neurons. In equations (2.1) and (2.2) it should be noted that the inter-neuron communication only occurs when the output of the neuron is high. Let us now consider three neurons A, B, and C that are linearly arranged with B between A and C. For this example, only A is receiving an input stimulus. At $n = 0$, the A neuron pulses sending a large signal to B. At $n = 1$, B receives the large signal, pulses, and then sends a signal to both A and C. At $n = 2$, the A neuron still has a rather large threshold value and therefore the stimulus is

14   2 Theory of Digital Simulation

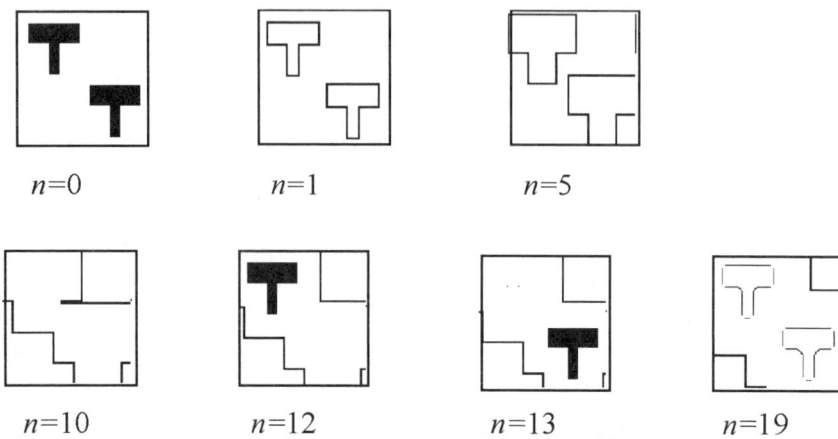

Fig. 2.3. A typical PCNN example

not large enough to pulse the neuron. Similarly, neuron B is turned off by its threshold. On the other hand, C has a low threshold value and will pulse. Thus, a pulse sequence progresses from A to C.

This process is the beginning of the *autowave* nature of the PCNN. Basically, when a neuron (or group of neurons) fires, an autowave emanates from that perimeter of the group. Autowaves are defined as normal propagating waves that do not reflect or refract. In other words, when two waves collide they do not pass through each other. Autowaves are being discovered in many aspects of nature and are creating a significant amount of scientific research [13, 23]. The PCNN, however, does not necessarily produce a pure autowave and alteration of some of the PCNN parameters can alter the behaviour of the waves.

Consider the image in Fig. 2.3. The original input consists of two 'T's. The intensity of each 'T' is constant, but the intensities of each 'T' differ slightly.

At $n = 0$ the neurons that receive stimulus from either of the 'T's will pulse in step $n = 1$ (denoted as black). As the iterations progress, the autowaves emanate from the original pulse regions. At $n = 10$ it is seen that the two waves did not pass through each other. At $n = 12$ the more intense 'T' again pulses.

The network also exhibits some synchronising behaviour. In the early iterations segments tend to pulse together. However, as the iterations progress, the segments tend to *de-synchronise*. Synchronicity occurs by a *pulse capture*. This occurs when one neuron is close to pulsing ($U < \Theta$) and its neighbour fires. The input from the neighbour will provide an additional input to U thus allowing the neuron to fire prematurely. The two neurons, in a sense, synchronise due to their linking communications. This is a strong point of the PCNN.

The de-synchronisation occurs in more complex images due to residual signals. As the network progresses the neurons begin to receive information indirectly from other non-neighbouring neurons. This alters their behaviour and the synchronicity begins to fail. The beginning of this failure can be seen by comparing $n = 1$ to $n = 19$ in Fig. 2.3. Note that the corners of the 'T' autowave are missing in $n = 19$. This phenomenon is more noticeable in more complicated images.

Gernster [14] argues that the lack of noise in such a system is responsible for the de-synchronisation. However, experiments shown in Chap. 3 specifically show the PCNN architecture does not exhibit this link. Synchronisation has been explored more thoroughly for similar integrate and fire models [22].

The PCNN has many parameters that can be altered to adjust its behaviour. The (global) linking strength, β, in particular, has many interesting properties (in particular effects on segmentation), which warrants its own chapter. While this parameter, together with the two weight matrices, scales the feeding and linking inputs, the three potentials, V, scale the internal signals. Finally, the time constants and the offset parameter of the firing threshold are used to adjust the conversions between pulses and magnitudes.

The dimension of the convolution kernel directly affects the speed that the autowave travels. The dimension of the kernel allows the neurons to communicate with neurons farther away and thus allows the autowave to advance farther in each iteration.

The pulse behaviour of a single neuron is greatly affected by α_Θ and V_Θ. The α_Θ affects the decay of the threshold value and the V_Θ affects the height of the threshold increase after the neuron pulses. It is quite possible to force the neuron to enter into a multiple pulse regime. In this scenario the neuron pulses in consecutive iterations.

The autowave created by the PCNN is greatly affected by V_F. Setting V_F to 0 prevents the autowave from entering any region in which the stimulus is also 0. There is a range of V_F values that allows the autowave to travel but only for a limited distance.

There are also architectural changes that can alter the PCNN behaviour. One such alteration is *quantized linking* where the linking values are either 1 or 0 depending on a local condition. In this system the Linking field is computed by

$$L_{ij}[n] = \begin{cases} 1 & \text{if } \sum_{ij} w_{ijkl} Y_{kl} > \gamma \\ 0 & \text{Otherwise} \end{cases}. \tag{2.6}$$

Quantized linking tends to keep the autowaves clean. In the previous system autowaves travelling along a wide channel have been observed to decay about the edges. In other words a wave front tends to lose its shape near its outer boundaries. Quantized linking has been observed to maintain the wavefronts shape.

Another alteration is called *fast linking*. This allows the linking waves to travel faster than the feeding waves. It basically iterates the linking and internal activity equations until the system stabilises. A detailed description will be discussed shortly. This system is useful in keeping the synchronisation of the system intact.

Finally, the initial values of Θ need to be discussed. If they are initially 0 then any neuron receiving a stimulus will cause the neuron to pulse. In a 'real world' image generally all of the neurons receive some stimulus and thus in the initial iteration all neurons will pulse. Then it will take several iterations before they can pulse again. From an image processing perspective the first few iterations are unimportant since all neurons pulse in the first iteration and then non pulse for the next several iterations. An alternative is to initially set the threshold values higher. The first few iterations may not produce any pulses since the thresholds now need to decay. However, the frames with useful information will be produced in earlier iterations than in the 'initially 0' scenario. Parodi [11] suggests that the Θ be reset after a few iterations to prevent de-synchronisation.

2.1.2 Time Signatures

The early work of Johnson [16] was concerned with converting the pulse images to a single vector of information. This vector, G, has been called the 'time signal' and is computed by,

$$G[n] = \sum_{ij} Y_{ij}[n]. \tag{2.7}$$

This time signal was shown to have an invariant nature with regard to alterations of the input image. For example, consider the two images in Fig. 2.4. These images are of a 'T' and a '+'.

Each image was presented to the PCNN and each produced a time signal, G_T and G_+, respectively. These are shown in Fig. 2.5.

Johnson showed that the time signal produces a cycle of activity in which each neuron pulses once during the cycle. The two plots in Fig. 2.5 depict single cycles of the 'T' and the '+'. As time progressed the pattern within the cycle stabilised for these simple images. The content of the image could be identified simply by examining a very short segment of the time signal - a single stationary cycle. Furthermore, this signal was invariant to large changes in rotation, scale, shift, or skew of the input object. Figure 2.6 shows several cycles of a slightly more complicated input and how the peaks vary with scaling and rotation as well as intensities in the input image. However, note that the distances between the peaks remain constant, providing a fingerprint of the actual figure. Furthermore, the peak intensities could possibly be used to obtain information on scale and angle.

2.1 The Pulse-Coupled Neural Network   17

Fig. 2.4. Images of a 'T' and a '+'

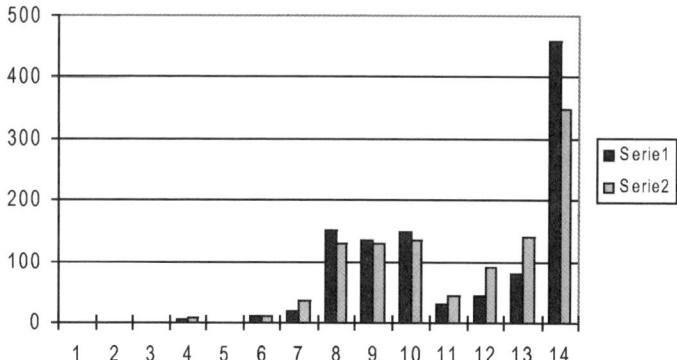

Fig. 2.5. A Plot of G_T (series 1) and G_+ (series 2) in arbitrary units (*vertical axis*). The *horizontal axis* shows the frame number and the *vertical axis* the values of G

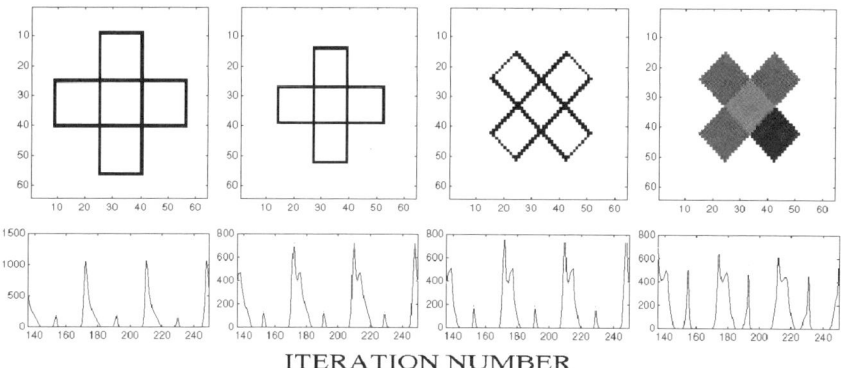

ITERATION NUMBER

Fig. 2.6. Plot of G for a slightly more complicated cross than in Fig. 2.5. The cross is then scaled and rotated and filled with shades of grey to show what happens to the time series

However, this only held true for these simple objects with no noise or background. Extracting a similarly useful time signal for "real-world" images has not yet been shown.

2.1.3 The Neural Connections

The PCNN contains two convolution kernels M and W. The original Eckhorn model used a Gaussian type of interconnections, but when the PCNN is applied to image processing these interconnections are available to the user for altering the behaviour of the network.

The few examples shown here all use local interconnections. It is possible to use long range interconnections but two impositions arise. The first is that the computational load is directly dependent upon the number of interconnections. The second is that PCNN tests to date have not provided any meaningful results using long range interconnections, although long range inhibitory connections of similar models have been proposed in similar cortical models [24].

Subsequent experiments replaced the interconnect pattern with a target pattern in the hope that on-target neurons would pulse more frequently. The matrices M and W were similar to the intensity pattern of a target object. In actuality there was very little difference in the output from this system than from the original PCNN. Further investigations revealed the reason for this. Positive interconnections tend to smooth the image and longer-range connections provide even more smoothing. The internal activity of the neuron may be quite altered by a change in interconnections. However, much of this change is nullified since the internal activity is compared to a dynamic threshold. The amount by which the internal activity surpasses the dynamic threshold is not important and thus the effects of longer-range interconnections are reduced.

Manipulations of a small number of interconnections do, however, provide drastic changes in the PCNN. A few examples of these are shown.

For these examples we use the input shown in Fig. 2.7. This input is a set of two 'T's.

The first example computes the convolution kernel by

$$K_{ij} = \begin{cases} 0 & \text{if } i = m \text{ and } j = m \\ 1/r & \text{Otherwise} \end{cases}, \quad (2.8)$$

where r is the distance from the centre element to element ij, and m is half of the linear dimension of K. In this test K was 5×5. Computationally, the feeding and linking equations are

$$F_{ij}[n] = e^{-\alpha_F \delta n} F_{ij}[n-1] + S_{ij} + (K \otimes Y)_{ij}, \quad (2.9)$$

and

$$L_{ij}[n] = e^{-\alpha_L \delta n} L_{ij}[n-1] + (K \otimes Y)_{ij}. \quad (2.10)$$

The resultant outputs of the PCNN are shown in Fig. 2.8.

The output first pulses all neurons receiving an input stimulus. Then autowaves are established that expand from the original pulsing neurons. These autowaves are two pixels wide since the kernel extends two elements

Fig. 2.7. An example of an image used as input

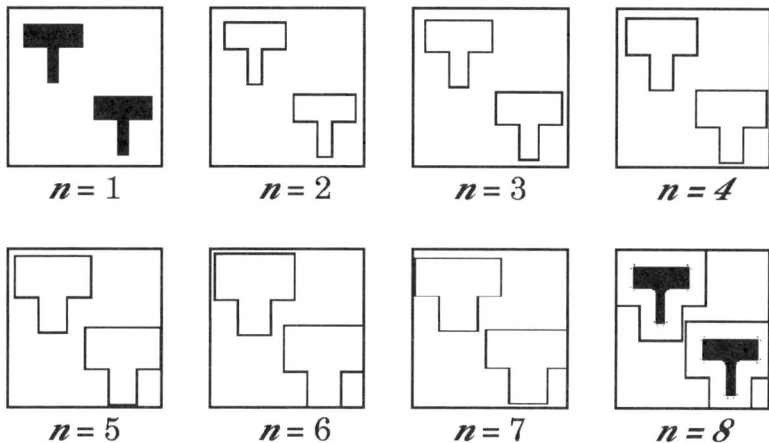

Fig. 2.8. Outputs of the PCNN

in any direction from the centre. These autowaves expand at the same speed in both vertical and horizontal dimensions again due to the symmetry of the kernel.

Setting the elements of the previous kernel to zero for $i = 0$ and $i = 4$ defines a kernel that is asymmetric. This kernel will cause the autowaves to behave in a slightly different fashion. The results from these tests are shown in Fig. 2.9.

The autowave in the vertical direction now travels at half the speed of the one in the horizontal direction. Also the second pulse of the neurons receiving stimulus is delayed a frame. This delay is due to the fact that these neurons were receiving less stimulus from their neighbours. Increasing the values in K could eliminate the delay.

The final test involves altering the original kernel by simply requiring that

$$K_{ij} = \begin{cases} K_{ij} & \text{if } i = m \text{ and } j = m \\ -K_{ij} & \text{Otherwise} \end{cases}. \tag{2.11}$$

20   2 Theory of Digital Simulation

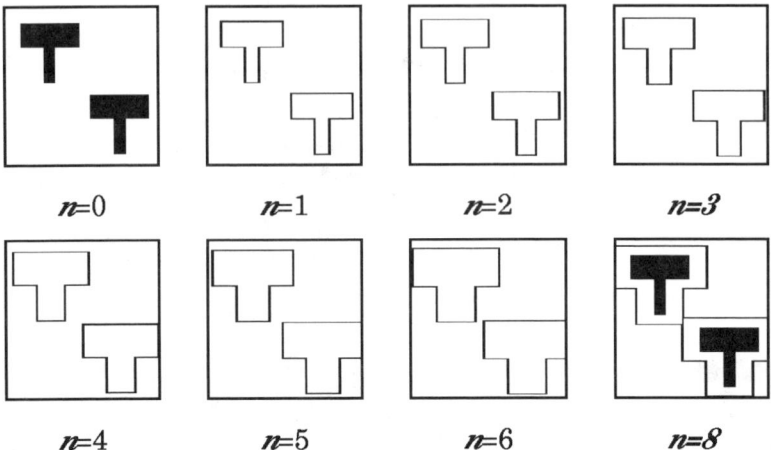

Fig. 2.9. Outputs of a PCNN with an asymmetric kernel, as discussed in the text. These outputs should be compared to those shown in Fig. 2.10

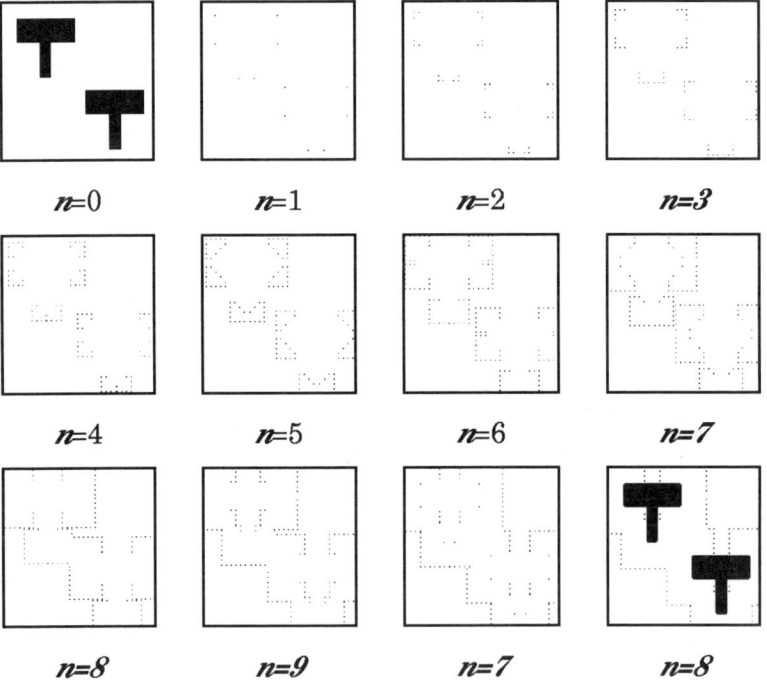

Fig. 2.10. Outputs of a PCNN with an on-centre/off-surround kernel

The kernel now has a positive value at the centre and negative values surrounding it. This configuration is termed *On-Centre/Off-Surround*. Such configurations of interconnections have been observed in the eye. Furthermore, convolutions with a zero-mean version of this function are quite often used as an "edge enhancer". Employing this type of function in the PCNN has a very dramatic effect on the outputs as is shown in Fig. 2.10.

The autowaves created by this system are now *dotted* lines. This is due to *competition* amongst the neurons since each neuron is now receiving both positive and negative inputs.

2.1.4 Fast Linking

The PCNN is a digital version of an analogue process and this quantisation of time does have a detrimental effect. Fast linking was originally installed to overcome some of the effects of time quantisation and has been discussed by [21] and [17]. This process allows the linking wave to progress a lot faster than the feeding wave. Basically, the linking is allowed to propagate through the entire image for each iteration.

Fast linking iterates the L, U, and Y equations until Y become static. The equations for this system are

$$F_{ij}[n] = e^{\alpha_F \delta n} F_{ij}[n-1] + S_{ij} + V_F \sum_{kl} M_{ijkl} Y_{kl}[n-1], \qquad (2.12)$$

$$L_{ij}[n] = e^{\alpha_L \delta n} L_{ij}[n-1] + V_L \sum_{kl} W_{ijkl} Y_{kl}[n-1], \qquad (2.13)$$

$$U_{ij}[n] = F_{ij}[n]\{1 + \beta L_{ij}[n]\}, \qquad (2.14)$$

$$Y_{ij}[n] = \begin{cases} 1 & \text{if } U_{ij}[n] > \Theta_{ij}[n-1] \\ 0 & \text{Otherwise} \end{cases}, \qquad (2.15)$$

REPEAT

$$L_{ij}[n] = V_L \sum_{kl} W_{ijkl} Y_{kl}[n-1], \qquad (2.16)$$

$$U_{ij}[n] = F_{ij}[n]\{1 + \beta L_{ij}[n]\}, \qquad (2.17)$$

$$Y_{ij}[n] = \begin{cases} 1 & \text{if } U_{ij}[n] > \Theta_{ij}[n-1] \\ 0 & \text{Otherwise} \end{cases}, \qquad (2.18)$$

UNTIL Y DOES NOT CHANGE

$$\Theta_{ij}[n] = e^{\alpha_\Theta \delta n} \Theta_{ij}[n-1] + V_\Theta Y_{ij}[n]. \qquad (2.19)$$

This system allows the autowaves to fully propagate during each iteration. In the previous system the progression of the autowaves was restricted by the radius of the convolution kernel.

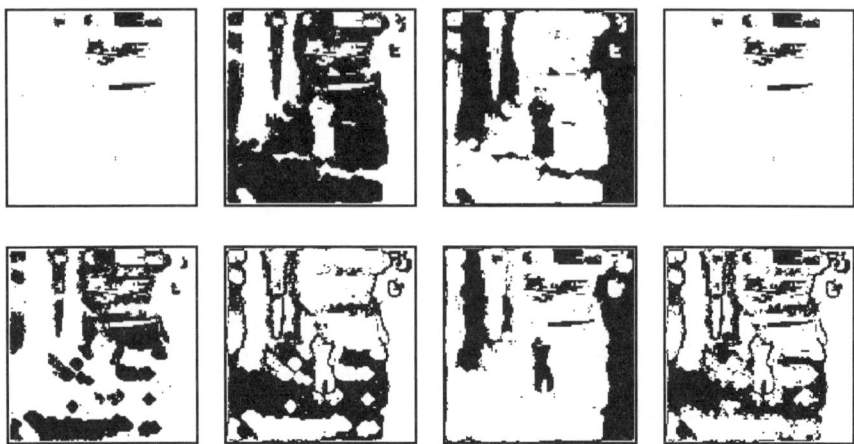

Fig. 2.11. Outputs of a fast-linking PCNN with random initial thresholds with the black pixels indicating which neurons have pulsed

Figure 2.11 displays the results of a PCNN with random initial threshold values. As can be seen, the fast linking method is a tremendously powerful method of reducing noise. It also prevents the network from experiencing segmentation decay. This latter effect may be desired if only segmentation was needed from the PCNN outputs and detrimental if the texture segmentation was desired.

2.1.5 Fast Smoothing

Perhaps the fastest way to compute the PCNN iterations is to replace both M and W with a smoothing operation. While this doesn't exactly match the theory it does offer a significant savings in computation time.

Consider the task of smoothing a vector $\boldsymbol{v}$. The brute force method of smoothing this vector is

$$a_j = \frac{1}{2\varepsilon+1} \sum_{i=j-\varepsilon}^{j+\varepsilon} \boldsymbol{v}_i. \qquad (2.20)$$

Each element in the answer a is the average over a short window of the elements in $\boldsymbol{v}$. The range of the window is determined by the constant ε. This equation is valid except for the ε elements at each end of a. Here the number of elements available for the averaging changes and the equation is adjusted according. For example, consider $j = 0$; there are no elements in the range $j - \varepsilon$ to 0. Thus, there are fewer elements for summation.

Consider now two elements of a that are not near the ends, $a_k = (\boldsymbol{v}_{k-\varepsilon} + \boldsymbol{v}_{k-\varepsilon} + 1 + \ldots + \boldsymbol{v}_{k+\varepsilon} - 1 + \boldsymbol{v}\boldsymbol{v}_{k+\varepsilon})/N$ and its neighbour $a_{k+1} = (\boldsymbol{v}_{k-\varepsilon} + 1 + \boldsymbol{v}_{k-\varepsilon} + 2 + \ldots + \boldsymbol{v}_{k+\varepsilon} + \boldsymbol{v}_{k+\varepsilon} + 1)/N$, where N is the normalization factor.

The only difference between the two is that a_{k+1} does not have $\boldsymbol{v}_{k-\varepsilon}$ and it does contain $\boldsymbol{v}_{k+\varepsilon} + 1$. Obviously,

$$a_{k+1} = a_k - \frac{(\boldsymbol{v}_{k+\varepsilon+1} - \boldsymbol{v}_{k-\varepsilon})}{N}. \tag{2.21}$$

Using this recursion dramatically reduces the computational load and it is more effective for larger ε. Thus, using this fast smoothing function reduces the computational load in generating PCNN results.

2.1.6 Analogue Time Simulation

As stated earlier the PCNN is a simulation in discrete time of a system that operates in analogue time. This is due solely to the ease of computation in discrete time. It is possible to more closely emulate an analogue time system. Computationally, this is performed by keeping a table of events. These events include the time in which each neuron is scheduled to pulse and when each inter-neural communication reaches its destination. This table is sorted according by the scheduled time of each event.

The system operates by considering the next event in the table. This event is computed and it either fires a neuron or modifies the state of a neuron because a communication from another neuron has reached this destination. All other events that are affected by this event are updated. For example, if a communication reaches its destination then it will alter the time that the neuron is predicted to pulse next. Also new events are added to the table. For example, if a neuron pulses then it will generate new communications that will eventually reach their destinations.

More formally, the system is defined by a new set of equations. The stimulus is U and it is updated via,

$$U(t + \mathrm{d}t) = e^{-\mathrm{d}t/\tau_U} U(t) + \beta U(t) \otimes K \tag{2.22}$$

where K defines the inter-neural communications and β is an input scaling factor. The neurons fire when a nonlinear condition is met,

$$Y_{ij}(t + \mathrm{d}t) = \begin{cases} 1 & \text{if } (\beta U(t) \otimes K)_{ij} > \Theta_{ij}(t) \\ 0 & \text{Otherwise} \end{cases}, \tag{2.23}$$

and the threshold is updated by,

$$\Theta(t + \mathrm{d}t) = e^{-\mathrm{d}t/\tau_\Theta} \Theta(t) + \gamma Y(t). \tag{2.24}$$

The effect is actually an improvement over the digital system, but the computational costs are significant. Figure 2.12 displays an input and the neural pulses. In order to display the pulses it is necessary to collect the pulses over a finite period of time, so even though they are displayed together the pulses in each frame could occur at slightly different times.

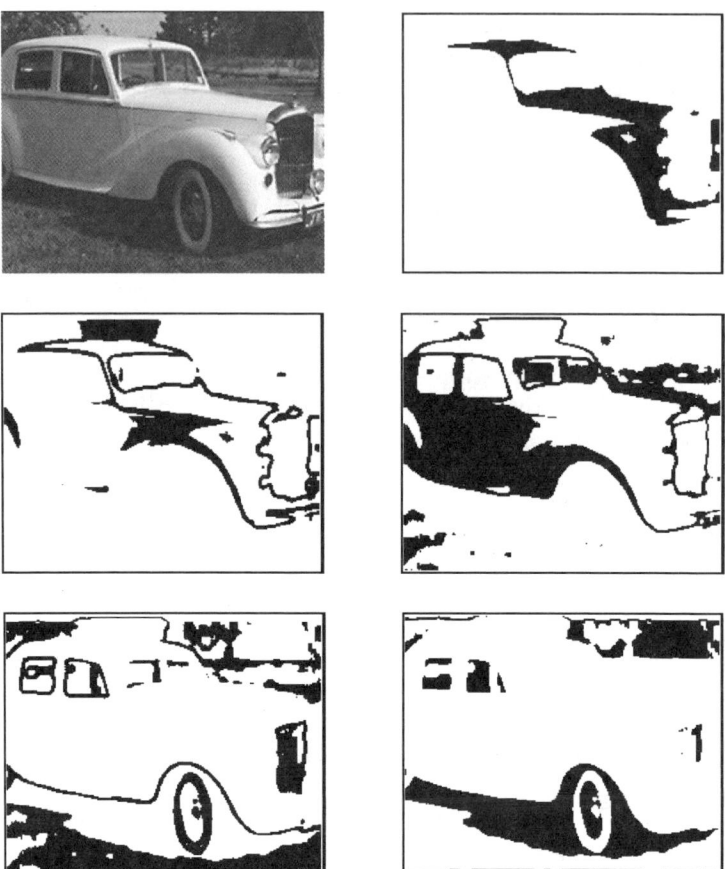

Fig. 2.12. An original image and collections of neural pulses over finite time windows

2.2 The ICM – A Generalized Digital Model

The PCNN is a digital model based upon a single biological model. As stated earlier there are several biological models that have been proposed. These models are mathematically similar to the Fitzhugh–Nagumo system in that each neuron consists of coupled oscillators. When the goal is to create image processing applications it is no longer necessary to exactly replicate the biological system. The important contribution of the cortical model is to extract information from the image and there is little concern as to the deviation from any single biological model.

The ICM is a model that attempts to minimize the cost of calculation but maintain the effectiveness of the cortical model when applied to images. Its foundation is based on the common elements of several biological models.

2.2.1 Minimum Requirements

Each neuron must contain at least two coupled oscillators, connections to other neurons, and a nonlinear operation that determines decisively when a neuron pulses. In order to build a system that minimizes the computation it must first be determined which operation creates the highest cost. In the case of the PCNN almost all of the cost of computation stems from the interconnection of the neurons. In many implementations users set $M = W$ which would cut the computational needs in half.

One method of reducing the costs of computation is to make an efficient algorithm. Such a reduction was presented in Sect.2.1.5 in which a smoothing operation replaced the traditional Gaussian type connections.

Another method is to reduce the number of connections. What is the minimum number of neurons required to make an operable system? This question is answered by building a minimal system and then determining if it created autowave communications between the neurons [18]. Consider the input image in Fig. 2.13 which contains two basic shapes.

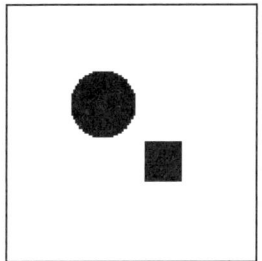

Fig. 2.13. An input image

The system that is developed must create autowaves that emanate from these two shapes. So, a model was created that connected each neuron to P other neurons. Each neuron was permanently connected to P random nearest neighbours and the simulation was allowed to run several iterations. The results in Fig. 2.14 display the results of three simulations. In the first $P = 1$ and the figure displays which neurons pulsed during the first 10 iterations. After 10 iterations this system stabilized. In other words the autowave stalled and did not expand. In the second test $P = 2$ and again the autowave did not expand. In both of these cases it is believed that the system had insufficient energy to propagate the communications between the neurons. The third test used $P = 3$ and the autowave propagated through the system, although due to the minimal number of connections this propagation was not uniform. In the image it is seen that the autowaves from the two objects did collide only when $P = 3$.

26   2 Theory of Digital Simulation

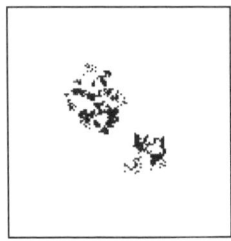  

Fig. 2.14. Neuron that fired in the first 10 iterations for systems with $P = 1$, $P = 2$, and $P = 3$

The conclusion is that at least three connections between neurons are needed in order to generate and autowave. However, for image processing applications the imperfect propagation should be avoided as it will artificially discriminate the importance of parts of the image over others.

Another desire is that the autowaves emanate as a circular wave front rather than a square front. If the system only contained 4 connections per neuron then the wave would propagate in the vertical and horizontal directions but not along the diagonals. The propagation from any solid shape would eventually become a square and this is not desired. Since the input image will be defined as a rectangular array of pixels the creation of a circular autowave will require more neural connections. This circular emanation can be created when each neuron is connected to two layers of nearest neighbours. Thus, $P = 24$ seems to be the minimal system.

2.2.2 The ICM

Thus, the minimal system now consists of two coupled oscillators, a small number of connections, and a nonlinear function. This system is described by the following three equations [19],

$$F_{i,j}[n+1] = fF_{i,j}[n] + S_{i,j} + W\{Y\}_{i,j}, \tag{2.25}$$

$$Y_{ij}[n+1] = \begin{cases} 1 \text{ if } F_{ij}[n+1] > \Theta_{ij}[n] \\ 0 \text{ Otherwise} \end{cases}, \tag{2.26}$$

and

$$\Theta_{i,j}[n+1] = g\Theta_{i,j}[n] + hY_{i,j}[n+1]. \tag{2.27}$$

Here the input array is S, the state of the neurons are F, the outputs are Y, and the dynamic threshold states are Θ. The scalars f and g are both less than 1.0 and $g < f$ is required to ensure that the threshold eventually falls below the state and the neuron pulses. The scalar h is a large value the dramatically increases the threshold when the neuron fires. The connections between the neurons are described by the function $W\{\}$ and for now these are still the $1/r$ type of connections. A typical example is show in Fig. 2.15.

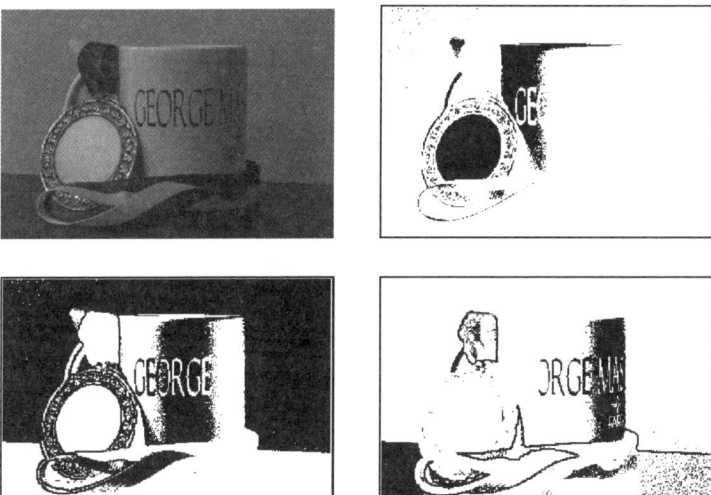

Fig. 2.15. An input image and a few of the pulse outputs from the ICM

Distinctly the segments inherent in the input image are displayed as pulses. This system behaves quite similar to the PCNN and is done so with simpler equations.

Comparisons of the PCNN and the ICM operating on the same input are shown in Figs. 2.16 and 2.17.

Certainly, the results do have some differences, but it must be remembered that the goal is to develop an image processing system. Thus, the results that are desired from these systems is the extraction of important image information. It is desired to have the pulse images display the segments, edges, and textures that are inherent in the input image.

2.2.3 Interference

Besides reducing the number of equations over the PCNN, the ICM has another distinct advantage. The connection function is quite different. The function $W\{\}$ was originally similar to the PCNN's M and W which were proportional to $1/r$. However, that model still posed a problem that plagued the PCNN. That problem was that of *interference*.

The problem of interference stems from the connection function $W\{\}$. Consider again the behaviour of communications when $W\{\} \sim 1/r$. In Fig. 2.18a there is an original image. The other images in Fig. 2.18 display the emanation of autowaves from the original object. This is also depicts how communications would travel if the ICM were stimulated by the original image.

These expanding autowaves are the root cause of interference. The autowaves expanding from non-target objects will alter the autowaves emanat-

28   2 Theory of Digital Simulation

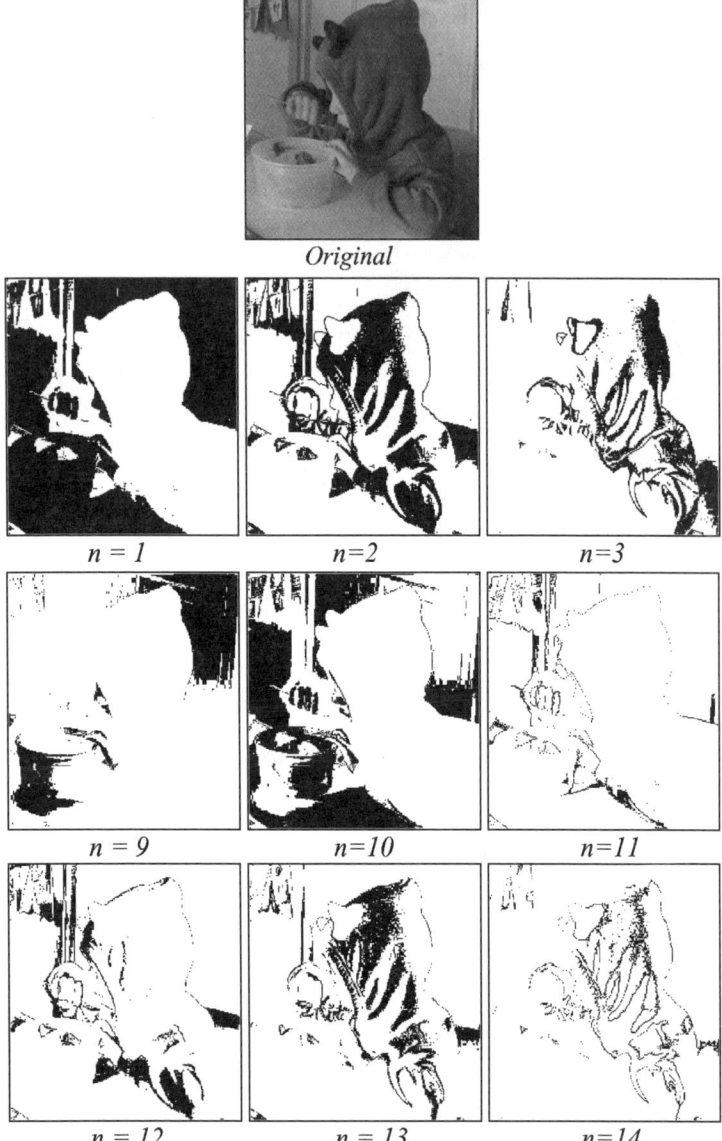

Fig. 2.16. An original image and several selected pulse images

ing from target objects. If the non-target object is brighter it will pulse earlier than the target object autowaves, and its autowave can pass through the target region before the target has a change to pulse. The values of the target neurons are drastically altered by the activity generated from non-target neu-

2.2 The ICM – A Generalized Digital Model

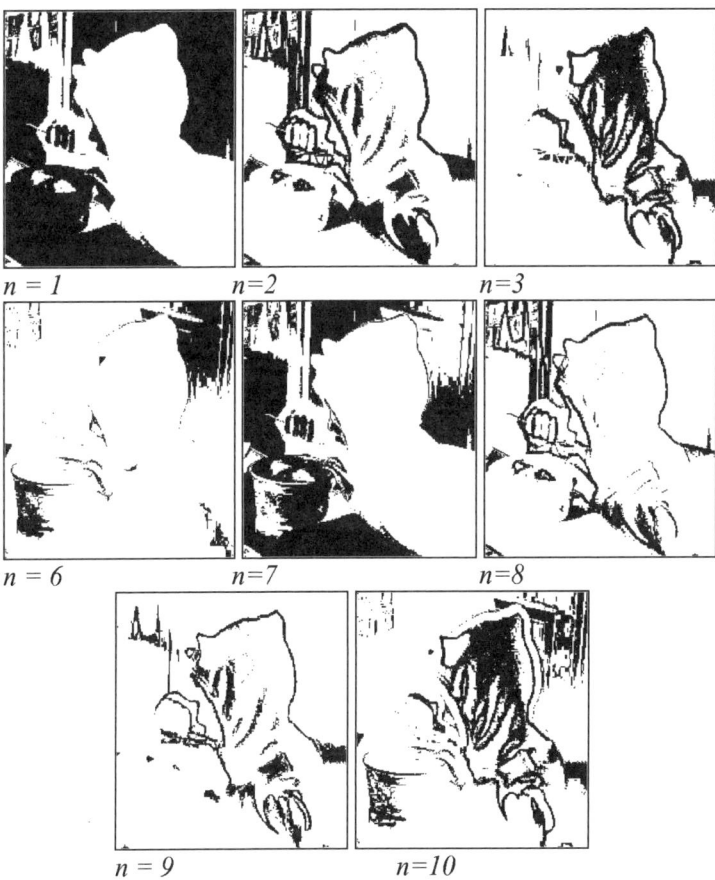

Fig. 2.17. Results from the ICM

rons. Thus, the pulsing behaviour of on-target pixels can be seriously altered by the presence of other objects.

An image was created by pasting a target (a flower) on a background (Fig. 2.19). The target was intentionally made to be darker than the background to amplify the interference effect. The ICM was run on both an image with the background and a image without the background. Only the pixels on-target were considered in creating the signatures shown in Fig. 2.20. The practice of including only on-target pixels is not possible for discrimination, but it does isolate the interference effects. Basically, the on-target pixels are altered significantly in the presence of a background. It would be quite difficult to recognize an object from the neural pulses if those pulses are so susceptible to the content of the background.

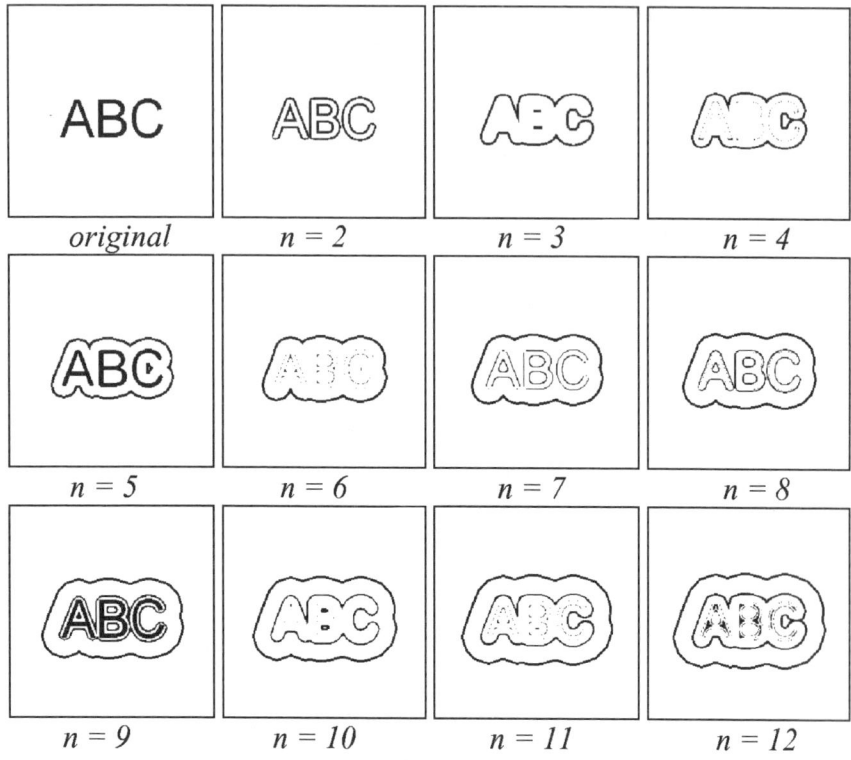

Fig. 2.18. Autowaves propagating from three initial objects. When the wavefronts collide they annihilate each other

Fig. 2.19. A target pasted on a background

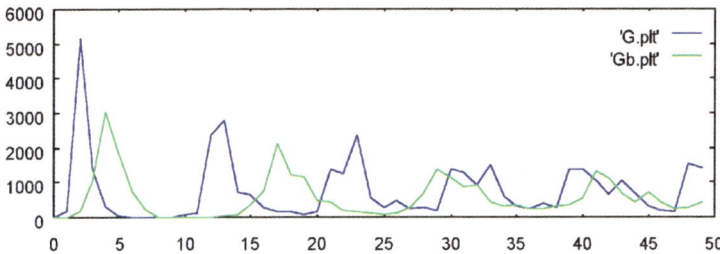

Fig. 2.20. The signature of the flower without a background (G.plt) and the signature of the flower with a background (Gb.plt)

2.2.4 Curvature Flow Models

The solution to the interference effect is based on *curvature flow* theory [20]. In this scenario the waves propagate towards the centripetal vectors that are perpendicular to the wave front. Basically, they propagate towards local centre of curvatures. For solid 2D objects the curvature flows will become a circle and then collapse to a point [15]. (There is an ongoing debate as to the validity of this statement in dimensions higher than two.)

Such propagation from Malladi is shown in Fig. 2.21. The initial frame presents a intricate 2D shape. This figure will eventually evolve into a circle and then collapse to a point. There is a strong similarity between this type of propagation and the propagation of autowaves. In both cases the wave front will evolve to a circle. The difference is that the autowaves will also expand the circumference with each iteration whereas the curvature flow will be about the same size as the original shape.

The interference in the ICM that lead to the deleterious behaviour in Fig. 2.20 was caused when the neural communications of one object interfered with the behaviour of another. In other words, the autowaves from the background infringed upon the territory owned by the flower. This stems from the ever expanding nature of the autowaves.

Curvature flow models evolve to the same shape as autowaves but do not have the ever-expanding quality. Thus, the next logical step is to modify the connection function $W\{\}$ to behave more like curvature flow wavefronts.

Fig. 2.21. The propagation of curvature flow boundaries

2.2.5 Centripetal Autowaves

A centripetal autowave follows the mechanics of curvature flow. When a segment pulses its autowave will propagate towards a circle and then collapse. It does not propagate outwards as does the traditional autowave. The advantageous result is that autowaves developed from two neighbouring objects will have far less interference.

The propagation of a curvature flow boundary is towards the local centre of curvature. The boundary, C, is a curve with a curvature vector κ. The evolution of the curve follows

$$\frac{\partial C}{\partial t} = \vec{\kappa} \cdot \hat{n}, \qquad (2.28)$$

where n is normal. In two-dimensional space all shapes become a circle and then collapse to a point. Such a progression is shown in Fig. 2.22 where a curve evolves to a circle and then collapses.

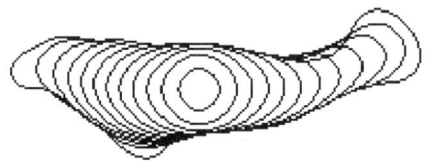

Fig. 2.22. The progression of an autowave from an initial shape

The ever-expanding nature of the autowaves leads to the interference and this quality is absent in a curvature flow model. Thus, the logical step is to modify the neural connections to behave as in the curvature flow model. This requires that the connections between the neurons be dependent upon the activation state of the surrounding neurons. However, in creating such connections the problem of interference is virtually eliminated. In this new scenario neural activity for on-target neurons is the same independent of the presence of other objects. This is a major requirement for the employment of these models as image recognition engines.

The new model will propagate the autowaves towards the local centre of curvature and thus obtain the name *centripetal autowaves*. The computation of these connections requires the re-definition of the function $W\{\}$.

Computations for curvature can be cumbersome for large images, so, an image-friendly approach is adopted. The curves in figure start with the larger, intricate curve and progress towards the circle and then collapse to a point. The neural communications will follow this type of curvature flow progression. Of course, in the ICM there are other influences such as the internal mechanics of the neurons which influence the evolution of the neural communications.

The function $W\{A\}$ is computed by

$$W\{A\} = A' = [[F_{2,A'}\{M\{A'\}\} + F_{1,A'}\{A'\}] < 0.5], \qquad (2.29)$$

where

2.2 The ICM – A Generalized Digital Model

$$A' = A + [F_{1,A}\{M\{A\}\} > 0.5] . \tag{2.30}$$

The function $M\{A\}$ is a smoothing function. The function $F_{1,A}\{X\}$ is a masking function that allows only the pixels originally OFF in A to survive as in,

$$[F_{1,A}\{X\}]_{ij} = \begin{cases} X_{ij} & \text{if } A_{ij} = 0 \\ 0 & \text{Otherwise} \end{cases}, \tag{2.31}$$

and likewise $F_{2,A}\{X\}$ is the opposing function,

$$[F_{2,A}\{X\}]_{ij} = \begin{cases} X_{ij} & \text{if } A_{ij} = 1 \\ 0 & \text{Otherwise} \end{cases}. \tag{2.32}$$

The operators $>$ and $<$ are threshold operators as in,

$$[X > d]_{ij} = \begin{cases} 1 & \text{if } x_{ij} \geq d \\ 0 & \text{Otherwise} \end{cases}, \tag{2.33}$$

and

$$[X > d]_{ij} = \begin{cases} 1 & \text{if } x_{ij} \leq d \\ 0 & \text{Otherwise} \end{cases}. \tag{2.34}$$

This system works by basically noting that a smoothed version of the original segment produces larger values in the off-pulse region and lower values in the on-pulse region in the same areas that the front is to propagate. The non-linear function isolates these desirable pixels and adjusts the communication wave front accordingly.

The centripetal autowave signatures of the same two images used to generate the results in Fig. 2.20 are shown in Fig. 2.23. It is easy to see that the background no longer interferes with the object signature. The behaviour of the on-target neurons is now almost independent of the other objects in the scene. This quality is necessary for image applications.

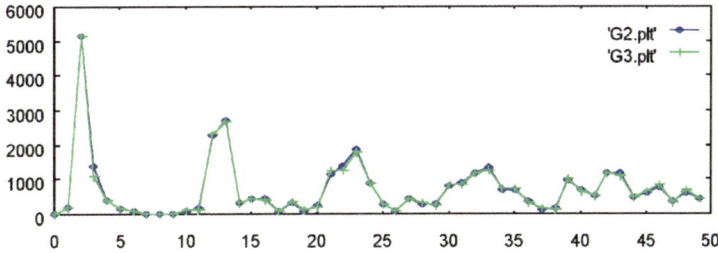

Fig. 2.23. The signatures of the flower and the flower with a background using the centripetal autowave model

2.3 Summary

Cortical models have been expressed in mathematical form for five decades now. The same basic premise of coupled oscillators or reaction-diffusion systems still apply to current models. Furthermore, in an image processing application the differences between the different models may not be that important. Therefore, speed and simplicity of implementation are more important here than replication of a biological system.

For image processing applications the model selected here is the ICM which consists of just three simple equations. Each neuron has two oscillators (the neuron potential and the neuron threshold) and each neuron has a nonlinear operation. Thus, when stimulated, each neuron is capable of producing a spike sequence, and groups of locally connected neurons have the ability to synchronize pulsing activity. When stimulated by an image is that these collectives can represent inherent segments of the stimulating image. Thus, a cortical model can become a powerful first step in many image processing applications.

The traditional neural connection schemes, however, allow neural communications to continually progress away from the originating region. While this may have some biological foundation, this property has been found to be deleterious to object recognition. Activity from one region can so drastically alter the activity in another region that object recognition becomes very difficult.

The solution to this problem is to alter the connections to the neurons so that they become sensitive to previous pulsing steps. In the model presented, these connections are described as centripetal autowaves such that the wave front progresses towards the local centre of curvature of the pulsing regions. This eliminates the ever-expanding nature of the waves without altering their shape-describing form.

The simplest applications of this ICM is to extract segments from images. A few examples were given though out the chapter to demonstrate the ability of the cortical models in image processing applications. This is only the beginning of the power that these algorithms can provide and the subsequent chapters will present more involved applications and results.

3 Automated Image Object Recognition

The development of the PCNN and ICM in the previous chapter was solely for the purpose of application to a variety of image processing and recognition tasks. In this chapter the PCNN and ICM will be used to directly extract pertinent information from a variety of images for the purpose of recognition.

Image recognition engines usually have multiple stages and often the first stage is to extract the information that is important to the recognition process. It could be argued that this is the most important stage, because the proper information is not extracted then the subsequent decision stage, no matter how powerful, will be unable to recognize the target. Furthermore, if the first stage can extract enough information then the decision stage could be a very simple algorithm. The extraction of enough information is basically the determining factor in the success of the recognition algorithm. In the following examples the PCNN and ICM are used to extract the important information for the individual application.

3.1 Important Image Features

Images have components that are important for the image-processing task. For example, in image recognition it is generally the edges, texture or segments of an image that are the most important features. Of course, this is strongly application dependent.

One traditional method of recognising objects within an image is through a Fourier filter. The logic of this type of filter is shown in Fig. 3.1. Basically, a filter containing the target centred in the frame is created and it is correlated with the input image. If the target exists in the input then a large correlation signal appears in the output correlation surface at the location of the target in the input image.

The Fourier filter system does have some serious drawbacks. First, if the target within the input scene does not exactly match the target image in the filter then the correlation signal is weaker. Thus, if the target were an aeroplane, which could be viewed at any angle with differing scale, and illumination, (perhaps obscured cloud), it is very difficult to design a filter that can recognise the target.

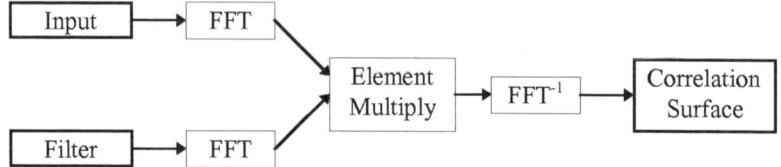

Fig. 3.1. Logical diagram of a Fourier filter

The point of this text though is not to discuss the pros and cons of the Fourier filter. It is, however, to indicate that this most popular system still relies heavily on the three main image components: edges, textures and segments.

In the Fourier space the lower frequencies of the image are located at the centre of the image and the higher frequencies are contained at the edges. Lower frequencies are present when the image has large areas of fairly uniform intensities. An aeroplane has lower frequencies from the hull and interior of the wings. Higher frequencies exist if the image has edges. For the aeroplane the edges of the craft and wings give rise to the higher frequencies.

There are two types of image recognition. One is generalisation in which the filter is designed to recognise a class of objects even though it may not be trained on the particular target in the input image. For example, a filter may be designed to recognise aeroplanes as opposed to helicopters. This filter would need the general shape of an aeroplane, which is contained in the lower frequencies. These lower frequencies are from the larger, more uniform areas of the image - in other words the segments.

Conversely, if the filter were designed to differentiate a particular type of aeroplane, from other flying vehicles then the general shape of an aeroplane is no longer useful. The more important information is contained in the higher frequencies (edges).

The Fourier filter works by matching the frequencies of the target with the frequencies in the input image. If the target exists in the input image then a strong match occurs and a large correlation signal is produced. Another manner in which to envision a Fourier filter is to take the target image and collocate it sequentially at every position in the input image. Eventually, the target in the filter and the target in the input are aligned and a match is easily seen. A Fourier filter is no more than performing a texture, segment, and edge match.

The whole point of this discussion is to indicate that the most common method of image recognition relies on the fundamentals of textures, edges and segments. Other types of image processing such as neural networks, morphology, and statistical processing also rely on these fundamentals.

While these methods are well understood theoretically, they have performed very poorly in the real world. Problems immediately arise when the training target(s) do not exactly match the input, which is quite often the case. The signal of the correlation drops and the noise from the background rises until the two become indistinguishable. The problems are far too complicated for these types of processors.

The PCNN/ICM models provide tremendous advantages here. First, the PCNN/ICM has an inherent ability to extract the fundamentals of the image. Second, the PCNN/ICM can simplify the image to allow recognition engines to perform a far easier task than is within their realm. The advantages of the PCNN/ICM in this form should not be a surprise since it is based on how mammals perform recognition.

The PCNN/ICM can extract the image fundamentals *inherently*. The PCNN/ICM does not need training or adjustments to extract these fundamentals from a wide range of images. Edges and segments are extracted at different iterations and segments can easily be seen over the course of a few iterations.

Segment extraction occurs since groups of neurons in a similar state tend to pulse in unison. Edges are extracted as the autowave expands from these segments. In the original form, the PCNN/ICM neurons will lose the unison pulsing according to the texture of the input. So, in time, the segments will tend to separate according to the texture.

The most important aspect of the PCNN/ICM performing these extractions is that it is an inherent quality of the PCNN/ICM. Traditional image processing has had engines that perform similar extractions but these are usually trained or designed to perform the task for particular applications. Furthermore, the PCNN/ICM will provide a higher quality of performance. For example, a popular method of extracting edges from an image is a Sobel filter, which consists of a small kernel in which the central elements are positive, and the surrounding elements are negative. Convoluting this kernel over the input image will result in an image with only the edge pixels 'on'. The problem is that this filter will produce a double line output for the edges that it sees. In other words, the edges are extracted but not cleanly. The PCNN/ICM extracts sharp, clean edges.

Also recall that the PCNN/ICM produces binary images. Segments that are extracted are shown as a solid uniform segment in the PCNN/ICM output. Edges are also of the same intensity in the output even though the input edges may have a gradient of intensities. These binary segments and edges are well organised in the Fourier plane. Performing recognition on these binary images is far easier to accomplish than performing recognition on the original input. This argument will be discussed in Sect. 3.7.

Thus, the PCNN/ICM algorithms are powerful pre-processors. They extract the fundamentals of an image (edges, textures, and segments) and can present far simpler binary images to recognition engines for analysis.

One of the major tools in image processing is the ability to extract segments from the image. This can lead directly to object identification algorithms as well as many other types of analysis.

38 3 Automated Image Object Recognition

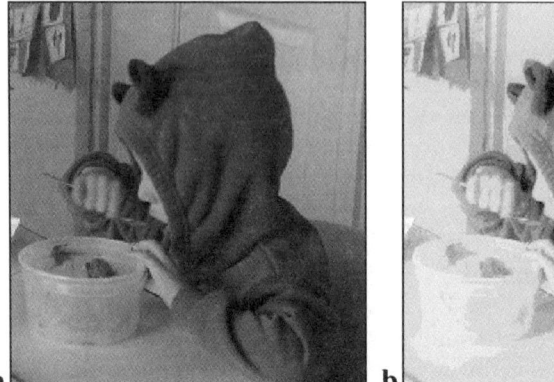

Fig. 3.2. The original image (**a**) and the weighted accumulation of pulses (**b**)

For this example, consider the ICM which has the ability to isolate the input image into its segments. One method that demonstrates the segmentation ability of the ICM is to accumulate the pulses weighted by their iteration. The output image is computed by,

$$P_{i,j} = \sum_n \alpha_n Y_{i,j}[n], \qquad (3.1)$$

α_n is a scaling factor that is inversely, monotonically proportional to n. An accumulation example is shown in Fig. 3.2.

In this image there are only a few gray levels in this image (one for each iteration over only the first cycle of pulses). Thus, segmentation of the image can be viewed in a collective context. If the ICM were a poor segmenter then the details of the original image would be lost in the cumulative pulse image. Small details such as those in the upper left corner, wrinkles in the fabric, the spoon, etc. are still quite distinct in the cumulative image, thus the segmentation is faithful to the segments inherent in the original image.

The PCNN and ICM also extract important edge information, although they are certainly not the first algorithms to do so. Edge extraction algorithms have been proposed for many years. Their purpose is to enhance the edges contained within an image and this enhancement is generally proportional to the sharpness of the edges.

There are two properties of the ICM pulses that make it ideal for edge extraction. The first is that the pulses are binary and thus the edges are sharp. That property is obvious. The second property is that the pulse segments are usually solid, and these assist in separating edges of objects from the edges due to texture.

The picture in Fig. 3.3 displays a skater with some notable properties. There are some very distinct edges, but there are also subtle edges due to texture such as the inside of his coat. There are also edges due between objects with similar grey scale values (i.e., the gloves and the coat sleeves).

3.1 Important Image Features   39

Fig. 3.3. An original image of the skater

One simple method of extracting edges from an image is to accumulate the differences between neighbouring pixels in both the vertical and horizontal direction,

$$a_{i,j} = \sqrt{\Delta_{x:i,j}^2 + \Delta_{y:i,j}^2}, \tag{3.2}$$

where $\Delta_{x:i,j}$ describes the change in values in the horizontal direction, as in,

$$\begin{aligned}\Delta_{x:i,j} &= \frac{(M_{i,j} - M_{i,j-1}) + (M_{i,j} - M_{i,j+1})}{2} \\ &= \frac{2M_{i,j} - M_{i,j-1} - M_{i,j+l}}{2},\end{aligned} \tag{3.3}$$

and

$$\Delta_{y:i,j} = \frac{2M_{i,j} - M_{i-1,j} - M_{i+1,j}}{2}. \tag{3.4}$$

The enhanced version of the image in Fig. 3.3 is shown in Fig. 3.4. To display this image in a print format it has been inverted such that the darkness of the lines indicates the sharpness of the edges. Subtle edges due to coat texture are detected but are too faint to see in the image.

Edge extraction with the ICM entails the demarcation of the edges from the pulse images. The level of edge detection (the intensity of the resultant edge) is inversely proportional to the ICM iteration number n,

$$b_{i,j} = \sum_{n=0}^{M} \beta_n Y_{i,j}[n], \tag{3.5}$$

The scalar β is the proportionality term and M is the number of iterations that are considered. The image in Fig. 3.5 displays the edged detection process with $M = 2$.

Fig. 3.4. An edge enhanced version of Fig. 3.3

Fig. 3.5. $M = 2$ edge detection   **Fig. 3.6.** $M = 6$ edge detection

These edges are similar to the strong edges in Fig. 3.4. Allowing M to increase produces more interesting results. As the ICM iterates segments will pulse again with de-synchronization. Segments that pulsed together in early iterations will tend to break apart in subsequent iterations. The image in Fig. 3.6 displays the process with $M = 6$.

The de-synchronization is strong enough through the higher texture regions (coat) that there are now many edges to display. Obviously, adjusting the value of M determines the types of edges that the ICM can extract.

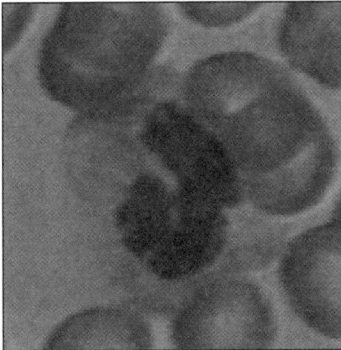

Fig. 3.7. An image of a red blood cell

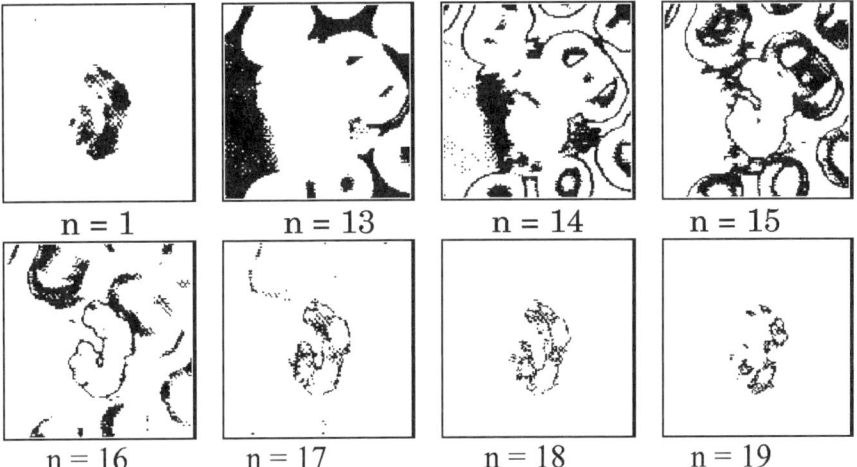

Fig. 3.8. Outputs of the PCNN with the pulsing pixels shown in black

3.2 Image Segmentation – A Red Blood Cell Example

Consider the image in Fig. 3.7, which is a red blood cell surrounded by white blood cells. The red blood cell has a dark nucleus and a cloudy cytoplasm.

This image was presented to a PCNN (in the original form) and it generated several outputs. Some of the outputs were trivial in that only a few of the pixels pulsed. The more interesting outputs are shown in Fig. 3.8.

At $n = 1$ the output has clearly segmented the nucleus of the red blood cell. At $n = 13$ the background is uniformly segmented. At $n = 14$ edges of the white blood cells and the cytoplasm of the red blood cell are seen. This is a very interesting output. In the original image the boundary between the background and the cytoplasm is very difficult to delineate, and finding this edge by traditional filtering would be extremely difficult. The PCNN, however, found this edge. At $n = 15$ and $n = 16$ edges of the nucleus and segments of the white blood cells are visible. Outputs $n = 17$ through $n = 19$ present the nucleus in three different frames. The de-synchronisation of the image at $n = 1$ is due to the texture of the nucleus.

In this one example it is easily seen that the PCNN has the ability to extract the fundamental features of an image. These extractions are presented in a binary format, which would be far easier to recognise than trying to determine the content of the image directly from the original input. In this fashion the PCNN has demonstrated that it can be a powerful pre-processor in an image recognition system.

3.3 Image Segmentation – A Mammography Example

Breast cancer is one of the leading causes of death for women the world over and its early detection is thus very important. In clinical examinations, physicians check for breast cancer by looking for abnormal skin thickenings, malignant tissues and micro calcifications. The latter are hard to detect be-

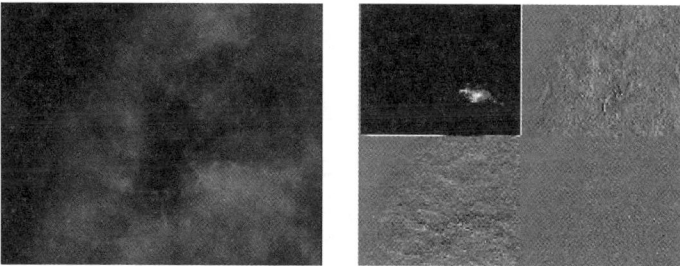

Fig. 3.9. A 2D Haar wavelet transform applied to an input image showing clusters of branching, pleomorphic calcification associated with poorly defined mass diagnosed as *duct carcinoma*

Fig. 3.10. The input to the PCNN (**left**) is a mammogram showing clusters of branching, pleomorphic calcification associated with poorly defined mass diagnosed as *duct carcinoma*. The pulsing pixels are shown in black

3.4 Image Recognition – An Aircraft Example

cause of similarity to normal glandular tissues. Wavelet transforms [25] have been used for automated processes has PCNN [30]. Examples of wavelet and PCNN processing of mammograms are shown in Figs. 3.9 and 3.10, respectively. The segmentation ability of the PCNN is clearly demonstrated in its ability to isolate these regions.

3.4 Image Recognition – An Aircraft Example

Consider the image in the upper left corner of Fig. 3.11. This image gives a grey scale input to a PCNN and the subsequent frames show the temporal series of outputs from a PCNN. Close to perfect edge detection is obtained and there are no problems identifying the aircraft e.g. from image number 3. Here the damaged wing tip is also easily seen.

However, this is not the case with a more complicated background such as mountains. Figure 3.12 shows a case where it is much harder to identify the aeroplane. A subsequent correlator is needed, and this will be discussed in Sect. 3.6.

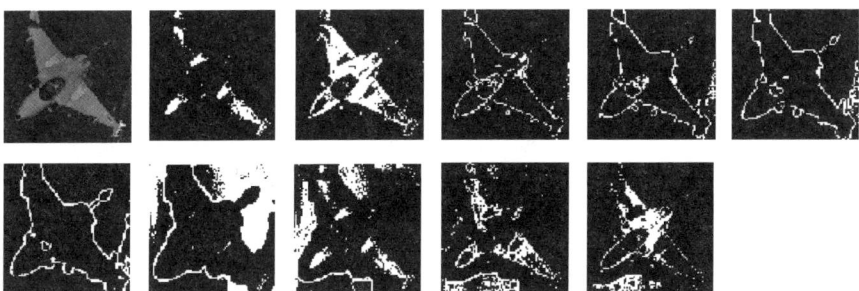

Fig. 3.11. The SAAB JAS 39 Gripen aircraft as an input to the PCNN. The initial sequence of temporal binary outputs are shown

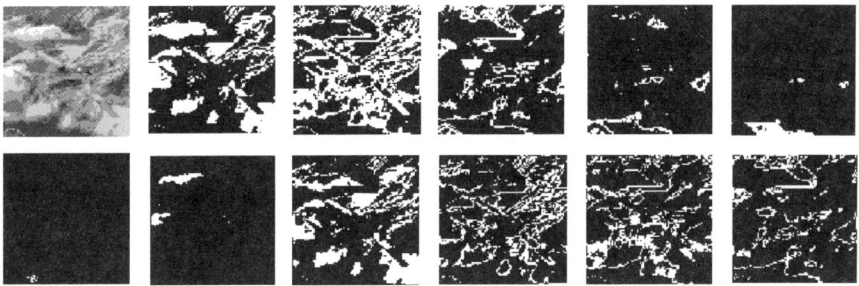

Fig. 3.12. The upper left image shows an aircraft as an input to the PCNN. However, this time an aeroplane is flying upside down in front of some Swiss Alps. It is hard to 'see' the aeroplane in the original input as well as in the temporal binary outputs

Table 3.1. Results of correct classification (in %)

	F5XJ		MIG-29		YF24		Learjet	
Net	Yes	No	Yes	No	Yes	No	Yes	No
LPN	95	86	90	85	86	87	92	87
BP	85	93	81	89	82	92	88	93
MD	82	78	83	90	72	90	90	88

The PCNN has been used as the first stage in a hybrid neural network [34] for Automatic Target Recognition (ATR). The time series provided a series of unique patterns of each of four aeroplanes.

(F5XJ, MIG-29, YF24 and the Learjet) used in this work. The input image to the ATR system was a 256 × 256 pixel image using 8 bit grey scale. The input data used for training the networks was obtained from simulated move-sequences of each aeroplane. The movie includes large variations in scale and 3D orientation of the aeroplanes. However, not all angles (and scale sizes) were included in the training data. This was done particularly in order to evaluate the generalisation capability of the system. Only the non-zero components of the first period of the PCNN 1D time series were used as input to the subsequent neural networks. The results [34] for several such 'conventional' neural network classifiers are shown in Table 3.1. The number of inputs was in all cases 43. Different neural networks were tested as final 'classificator' of PCNN output, The Logicon Projection Network^{TM}, LPN [35], the Back Propagation network, BP, The Radial Basis Function network, RBF, using two different algorithms, the Moody–Darken algorithm [32]. The numbers in Table 3.1 represent the mean value of correct classification of Yes/No, in %, for each of the four classes, together with the standard deviation, σ, of each class. The LPN and the BP get total average results that are nearly equal. However the LPN is always better classifying the signal (Yes), and BP is always better classifying the background (No). Moody–Darken network showed large standard deviations in several tests, especially for classifying the signal (Yes) for the F5XJ and YF24 aeroplanes.

3.5 Image Classification – Aurora Borealis Example

Auroras are spectacular and beautiful phenomena that occur in the auroral oval regions around the polar caps. Here geomagnetic field lines guide charged particles (mainly electrons and protons of magnetospheric or magnetosheath origin) down to ionospheric altitudes. When precipitating, the particles lose their energy via collisions with the neutral atmosphere, i.e. mainly oxygen and nitrogen atoms. At an altitude range of about 75–300 km, some of the atmospheric constituents can be excited to higher energy levels and this can lead to the formation of auroral light. The auroras during magnetospheric sub storms and, especially, the great auroras during magnetospheric storms

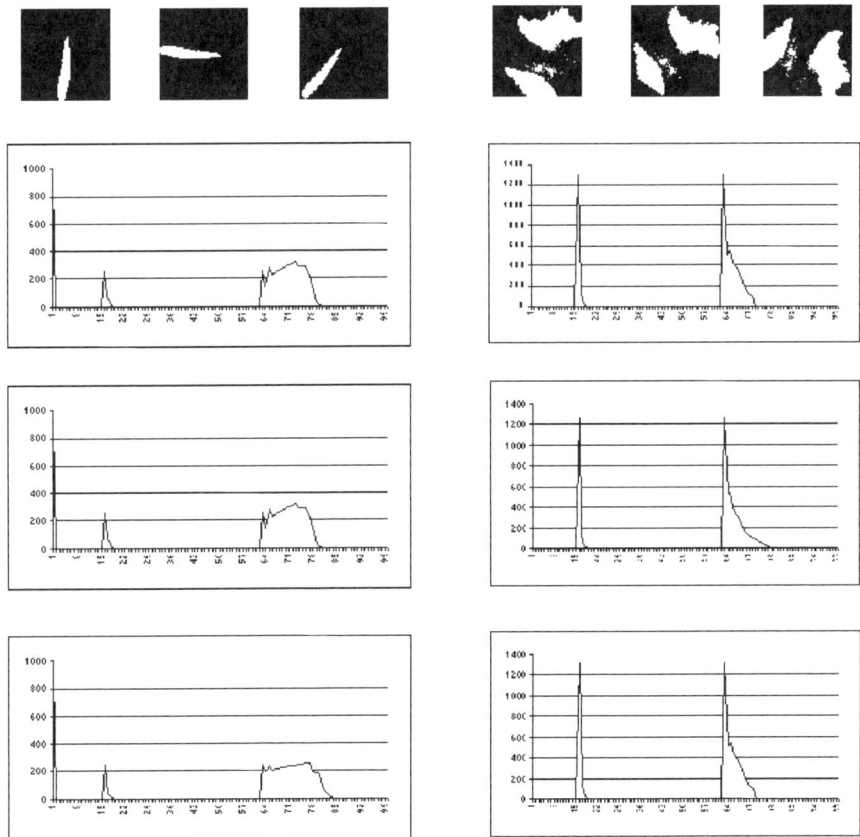

Fig. 3.13. Examples of Aurora Borealis and their resulting time signals when presented to a PCNN. A single arc (**left**) and a double arc (**right**) both retain their respective time signals when the images are rotated (**top**)

can create extremely impressive and beautiful spectacles. However, they are different, non-regular appearance makes it very difficult to design systems to recognise and categorise them.

Several Aurora projects include all-sky cameras inside automated observing stations like the ALIS ones in Sweden and those near Eagle, a small Alaskan village. In view of the large number of images it is important to make the classification of auroras automatic too. The classification is made is several different ways. The schemes may be based on where the auroras occur, how they look and/or on a physical model. In either case one needs to pre-process the images to take care of rotation, scale and translation effect. The PCNN has shown to be a good pre-processor in this case. Examples of this are shown in Fig. 3.13. By using the PCNN as a pre-processor, a subsequent neural pre-processor can be trained to identify the different classes of Auroras [36, 37].

3.6 The Fractional Power Filter

Many of the subsequent examples will use the PCNN or ICM as the information extraction stage and then a correlation filter to make the decision. This filter is the fractional power filter (FPF) [26].

The FPF is a composite filter that can trade-off generalisation and discrimination. The composite nature of the filter allows for invariance to be built into the filter. This is useful when the exact presentation of the target can not be predicted or the filter needs to be invariant to alterations such as rotation, scale, skew, illumination, etc. The FPF also allows the same filter to be used to detect several different targets.

While the composite nature of the filter is very desirable for this application, it means some trade-offs are unavoidable. The binary objects on which the filter operates, can, for example, be confused with other objects. An increase in the discrimination ability of the filter cures this problem.

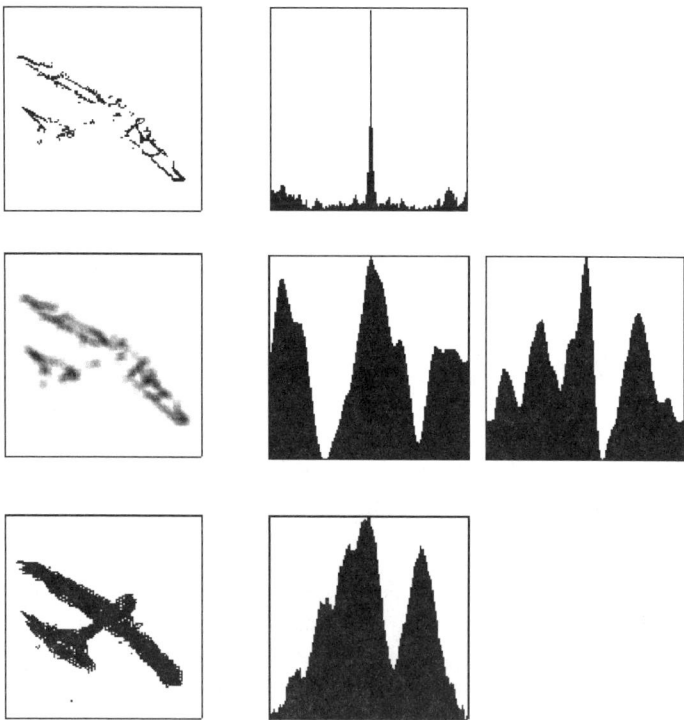

Fig. 3.14. Inputs with corresponding correlations. The plots on the right show a linear slice through the correlation surface through the target location

The FPF is a correlation filter built by creating a matrix $\hat{X}$ whose columns are the vectorized Fourier transform of the training images. The filter h is built by,

$$\hat{h} = D^{-1/2}\hat{Y}\left[\hat{Y}^T\hat{Y}\right]^{-1}\mathbf{c}, \tag{3.6}$$

where

$$\hat{Y} \equiv D^{-1/2}\hat{X}, \tag{3.7}$$

$$D_{ij} = \frac{\delta_{ij}}{N}\sum_k |\hat{\nu}_{k,i}|^p, \quad p = [0, 2], \tag{3.8}$$

and the vector $\mathbf{c}$ is the constraint vector.

A generalising FPF has $p = 0$, which is the synthetic discriminant filter. The fully discriminatory filter has $p = 2$, which is the minimum average correlation energy filter. A good review of this family of filters is found in [31]. Values of p between 0 and 2 trade-off generalisation and discrimination.

An example of the results from a PCNN with a FPF is shown in Fig. 3.14. Three different masks are shown together with a cut in the correlation matrix. The top row uses a 'hand-tailored' mask and yields a strong correlation. A more 'diffuse' mask (second and third row) will still yield strong correlations, in particular if several images are considered.

3.7 Target Recognition – Binary Correlations

One application of the FPF is to find targets in an image. In this case a single image is used for the recognition of the target and the fractional power term is adjusted to increase the importance of edge information. Consider the first image in Fig. 3.2 and the task of identifying his hand. The procedure would first mask the target region and place it in the centre of a frame as shown in Fig. 3.15.

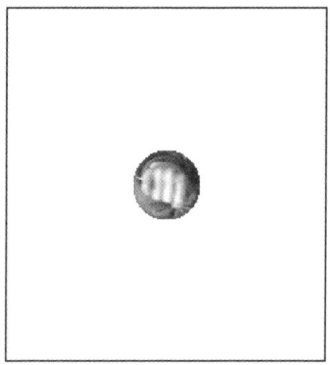

Fig. 3.15. The hand as an isolate target

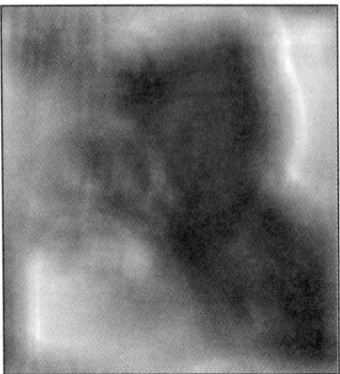 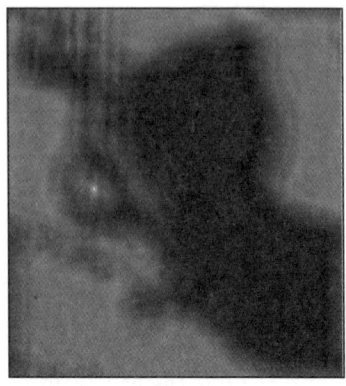

Fig. 3.16. The correlation with an FPF and $p = 0.3$

Fig. 3.17. The correlation of a pulse image ($n = 1$) with an FPF ($p = 0.3$)

This image can be correlated with the original image and the FPF will force a positive response at the location of the target. At the centre of the target in the original image the correlation value will be close to 1 since that was the value set in the constraint vector c. However, none of the other correlation surface values are similarly constrained except that as the value of the fractional power term is increased the overall energy of the correlation surface is reduced. This is a trade-off and as the fractional power term is increased the filter has an increasingly difficult time identifying anything that is not exactly like the training image. This discrimination also includes any windowing operation. Even though the hand in the filter is exactly like the hand in the original image the fact that the filter has a window about the target would detrimentally affect the filter response if the fractional power term became too high.

Thus, the trade-off between generalisation and discrimination can prevent a good solution from being achieved. Figure 3.16 displays the correlation between the image in Fig. 3.2a and the filter made from Fig. 3.15 with $p = 0.3$.

There is a small white dot at the centre of the target which indicates a strong correlation between the image and the filter. However, the filter also provides significant responses to other areas of the target. The conclusion is that the filter is generalising too much. If the fractional power term is increased to lower the generalisation then it also loses the ability to find the target.

By employing the ICM this problem is solved. The ICM will create several pulse images and in one the target will pulse. Complicated targets may their pulses spread over a few iterations, but in this case the hand is a simple target. The reason that this is important is that the pulse image presents the target as a solid object with sharp lines. This is much easier for a correlation filter to detect. The image in Fig. 3.17 displays the correlation of a pulse image and an FPF built from a binary version of the target.

3.7 Target Recognition – Binary Correlations   49

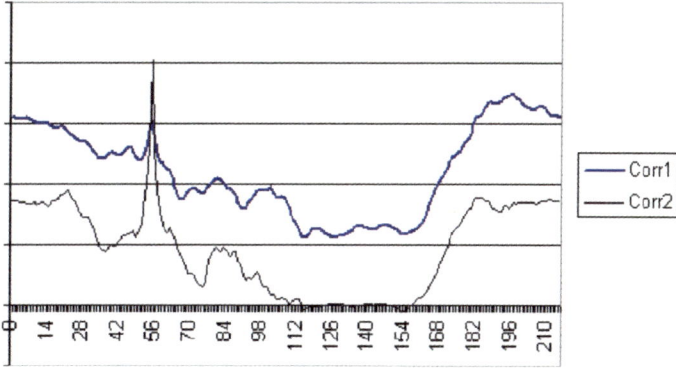

Fig. 3.18. Slices through the two correlation surfaces. Clearly in the second case the FPF can find the target from the ICM pulse image

The bright spot clearly indicates the detection of the target without interference from other objects in the image. To further demonstrate the detection ability using the ICM the image in Fig. 3.18 displays a slice through each correlation surface (Figs. 3.16 and 3.17) through the target pixel. These are the values of the pixels along the horizontal row in Fig. 3.17 that passes through the target. Clearly, the method using the ICM provides a much more detectable response.

The employment of the ICM in this case is not a subtle affect. One of the unfortunate features of a correlation filter is that it prefers to find targets with higher energy. Darker targets are harder to find. In the ICM the darker objects will eventually create a pulse segment and in that iteration the target will be the bright object making it easier for the filter to identify it. Consider the images in Fig. 3.19 in which there is an image with a young boy. In this case the boy is that target and the dark object. It would be different to build a filter to find the boy since he is so dark. Also in this figure are some of the pulse images. In the third pulse image the boy's pixels pulse. In this iteration he is now the high energy object and it much easier to find with a filter. For display purposes the neurons that pulse are shown in black.

Again, building an FPF from the pulse image the identification of the dark target is easily performed. The correlation is shown in Fig. 3.20 and the slice of the correlation surface through the target pixel is shown in Fig. 3.21. In this case the tree trunks pulsed in the same iteration and so they also correlate with the filter. Unfortunately, in this case the tree trunks are about the same width as the boy and so they provide strong correlations. However, the FPF has the ability to increase the discrimination and thus the correlation signal of the trees is not as sharp as that of the boy. The remainder of the process merely finds large, sharp peaks to indicate the presence of a target. The peak belonging to the boy in this case is definitely the sharpest peak and now a decision is possible.

50   3 Automated Image Object Recognition

Fig. 3.19. An original image and some of the pulse images. In one of the pulse images the target is the bright object and in another the outline of the target is the bright object

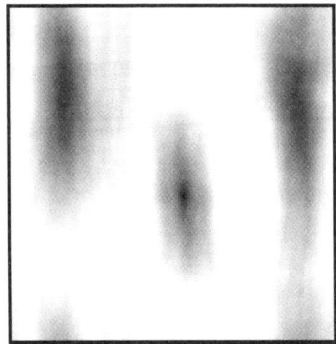

Fig. 3.20. The correlation response of the pulse image and an FPF. The darkest spot depicts the highest peak

It was the duty of the PCNN to extract the target and present it as a high energy collective pulse segment. A Fourier filter such as the FPF could easily detect the target and produce a sharp correlation signal. Whereas, a filter operating on the original image would have a very difficult time in finding the target.

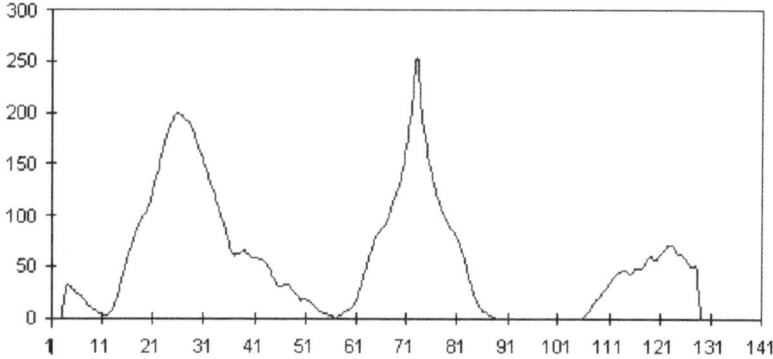

Fig. 3.21. A slice through the centre of the target of the correlation surface

3.8 Image Factorisation

The major problem in automatic target recognition is that images of the target change with scale, lighting, orientation, etc. One way to get around this problem has recently been suggested by Johnson [28, 33]. It involves a hierarchical image decomposition, which resolves an image into a set of image product factors. The set is ordered in scale from coarse to fine with respect to image detail. When the factored set is multiplied together it reproduces the original image. By selecting factors, coarse scene elements such as shadows, and fine scene factors such as noise, can be isolated. The scale of detail is controlled by the linking strength of a pulse coupled neural network, on which the system is based.

The factorisation system consists of three layers as shown in Fig. 3.22. The first layer is a PCNN and its purpose is to define the limit of detail to which the input will be resolved, both spatially and in intensity. The second layer serves to re-normalize the input to the third layer. The second layer is also a PCNN and together with the third layer it operates in a cyclic manner to provide the ordered output set of factors. The re-normalization is via a shunting action approximated by dividing the previous input to the third layer by the current output of the third layer. The output set consists of the outputs of the third layer in the order they were generated. Both PCNN types uses a single-pass, linear decay model with nearest-neighbour sigmoidal linking.

The algorithm is discussed in some detail in [28], but is simply expressed as,

$$G(n) = \frac{G(n-1)}{Y_2(n)}, \tag{3.9}$$

$$\beta(n) = k\beta(n-1), \tag{3.10}$$

where $G(0) = Y_1$ and $k < 1$.

Here Y_1 is the output of the first PCNN, $G(n)$ is the input to the second PCNN at the beginning of the nth cycle, $Y_2(n)$ is the output of the

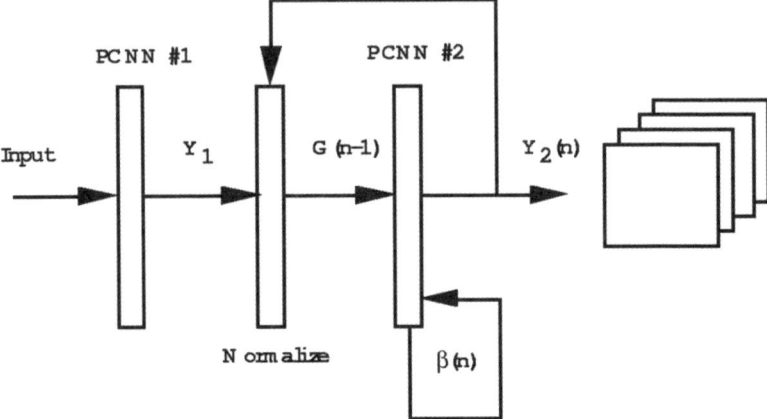

Fig. 3.22. Hierarchical image factor generation. An input image is decomposed into a set of factor images, ordered from coarse to fine in image detail. The product of the set is equal to the original image

second PCNN at the end of the $(n-1)$th cycle, and $k < 1$ is the linking strength reduction factor per cycle. In the above equation $\beta(0)$ is the initial value assigned by the operator. Together with the parameter k, it determines the initial coarseness resolution and the number of cycle's n. In the above, the spatial dependence of G and the PCNN output images Y_1 and Y_2 is suppressed, as the re-normalisation is applied on a pixel-by-pixel basis. The change of β is global, the same value being used by every pixel.

On the first cycle the second layer passes its input directly to the third layer. Its coarsely grey-scale quantifies it, giving an output that is coarse in both spatial and in intensity detail. When it is used by the second layer to normalise the original input, the new input, and all successive ones, will be between zero and one. As the second input is processed by the output PCNN, which now uses a reduced value of its linking strength, only the regions of intensity less than unity give values different than those of the first output.

3.9 A Feedback Pulse Image Generator

The PCNN can be a very powerful front-end processor for an image recognition system. This is not surprising since the PCNN is based on the biological version of a pre-processor. The PCNN has the ability to extract edge information, texture information, and to segment the image. This type of information is extremely useful for image recognition engines. The PCNN also has the advantage of being very generic. Very few changes (if any) to the PCNN are required to operate on different types of data. This is an advantage over previous image segmentation algorithms, which generally require information about the target before they are effective.

3.9 A Feedback Pulse Image Generator

There are three major mechanisms inherent in the PCNN. The first mechanism is a dynamic neural threshold. The threshold, here denoted by Θ, of each neuron significantly increases when the neuron fires, then the threshold level decays. When the threshold falls below the respective neuron's potential, the neuron again fires, which raises the threshold, Θ. This behaviour continues which creates a pulse stream for each neuron.

The second mechanism is caused by the local interconnections between the neurons. Neurons encourage their neighbours to fire only when they fire. Thus, if a group of neurons is close to firing, one neuron can trigger the entire group. Thus, similar segments of the image fire in unison. This creates the segmenting ability of the PCNN. The edges have different neighbouring activity than the interior of the object. Thus, the edges will still fire in unison, but will do so at different times than the interior segments. Thus, this algorithm isolates the edges.

The third mechanism occurs after several iterations. The groupings tend to break in time. This "break-up" or de-synchronisation is dependent on the texture within a segment. This is caused by minor differences that eventually propagate (in time) to alter the neural potentials. Thus, texture information becomes available.

The Feedback PCNN (FPCNN) sends the output information in an inhibitory fashion back to the input in a similar manner to the rat's olfactory system. The outputs are collected as a weighted time average, A, in a fashion similar to the computation of Θ except for the constant V,

$$A_{ij}[n] = e^{-\alpha_A \Delta t} A_{ij}[n-1] + V_A Y_{ij}[n], \qquad (3.11)$$

where V_A is much lower than V. In our case, $V_A = 1$. The input is then modified by,

$$S_{ij}[n] = \frac{S_{ij}[n-1]}{A_{ij}[n-1]}. \qquad (3.12)$$

The FPCNN iterates the PCNN equations with (3.11) and (3.12) inserted at the end of each iteration.

Two simple problems are shown to demonstrate the performance of the FPCNN. The first problem used a simple square as the input image. Figure 3.23 displays both the input stimulus S and the output Y for a few iterations until the input stabilised.

At $n = 5$ an output pulse has been generated by the FPCNN which is a square that is one pixel smaller than the original square on all four sides. At this point in the process the output of the FPCNN matches that of the PCNN. The PCNN would begin to separate the edges from the interior. In the case of the FPCNN, however, the input will now experience feedback shunting that is not uniform for the entire input. This is where the PCNN and the FPCNN differ.

As the iterations continue the activations go from a continuous square to just the edges of a square, and finally to just the four corners. The four

54   3 Automated Image Object Recognition

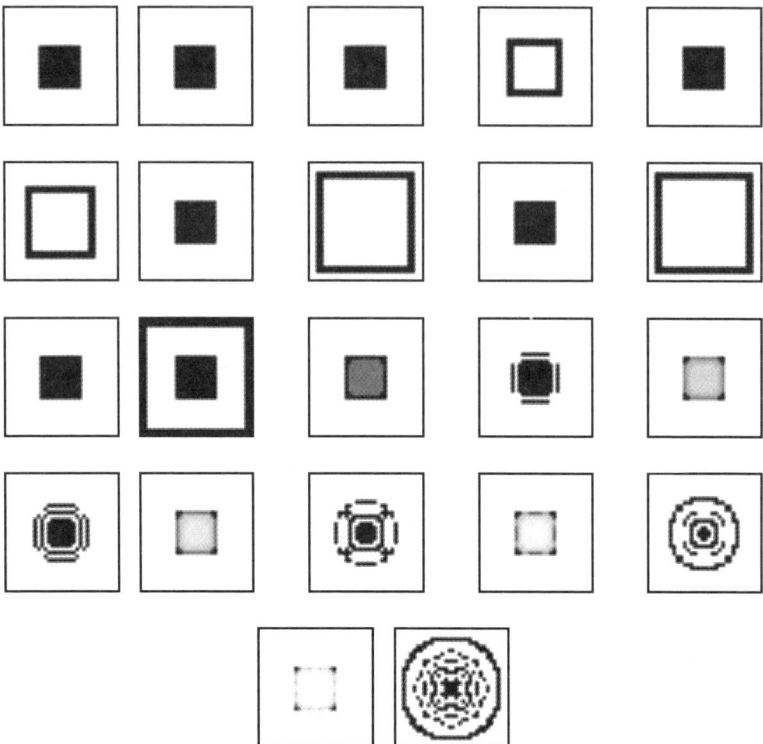

Fig. 3.23. Input and output pairs for FPCNN calculations for a solid square input

corners will remain on indefinitely. It is interesting to note that for the case of a solid triangle with unequal angles, the same type of behaviour occurs except that the corner with the smallest angle will dominate. The input goes from a solid triangle to an edge mostly triangle to the three corners to the dominant corner.

The second test to be shown is that of a square annulus. The results of this test are shown in Fig. 3.24. This process took more iterations than the solid square so only a few are shown. For the initial iterations the behaviour of the network mimics the solid square case. The edges become dominant. However, the steady state image is a bit more complicated than the four corners.

In order for the input to stabilise, a few conditions must be met. First, the input values must be contained above a threshold. If the input values fall too low, then the decay (α terms) will dominate and the input will eventually disappear. The second condition is that the output must completely enter into the multiple pulse realms. In this scenario all of the output elements are always on. This condition can be seen in the examples. When this occurs, the feedback and the interconnections become constant in time. In actuality, the outputs are independent variables and it is the outputs that force all other

Fig. 3.24. Input and output pairs for FPCNN calculations for a square annulus input

fluctuations within the network. When the outputs stabilise, there exists no other force within the network that can alter the input or any other network value.

3.10 Object Isolation

The Feedback PCNN (FPCNN) uses the FPF to filter the outputs of the PCNN. In turn, the FPF outputs are used to alter both the input [29] and the PCNN to isolate the objects in the input image and to enhance the PCNN's sensitivity to the masked areas in the FPF. The PCNN then extracts only a subset of segments or edges of the image. The benefit of all this is that segments and edges are very clean even when the input is quite confusing.

This attribute allows the filter to operate on clean segments and edges of the image. This is extremely important since the filter performance is dramatically improved when operating on clean images. Furthermore, the fact that the inputs to the filter are binary allows the filter to operate on sharp edges which is very important for discrimination.

The FPCNN logic flow is shown in Fig. 3.25. Initially, the dynamic input is a copy of the original input. The filters that are used in the FPF and the recursive image generator (RIG) are training targets. The intent of the system is to produce large correlation spikes where the target segment is

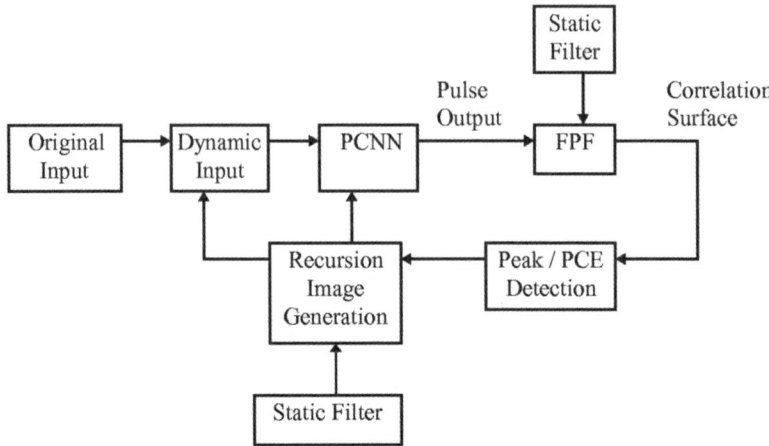

Fig. 3.25. The schematic of the feedback PCNN system

produced by the PCNN. However, other clutter may also produce significant correlation signals. Therefore, the correlation surface is examined for peaks.

In this system both the peak height and the peak to correlation energy (PCE) [27] are examined. It should be noted that neither has to be significantly better than the noise. In the example shown in the subsequent section an acceptable peak height is one-fourth of the training value. Not only do the filters operate on very clean inputs, but also their performance requirements for the FPCNN are not stringent.

The RIG generates an image that is the OR of binary target images that are centred about each sufficient correlation spike. This generated image is inverted and negatively alters the dynamic input. The non-inverted generated image is used as a reset mask for the PCNN. This latter step slows the natural tendency of the PCNN to separate the image segments in subsequent iterations.

Areas of the input image that did not provide significant correlation spikes are shunted. Consequently, only the targets in the input are enhanced. Strong correlations from non-target areas are allowed infrequently. These undesired correlations occur early in the iteration sequence and degrade with the other non-target areas as the FPCNN progresses.

Consider the image shown in Fig. 3.19a. This image contains a small boy (which will be the target) and several other objects acting as background clutter. The boy is certainly not the most intense object within the field of view. Furthermore, edges between the target and background are not the most prominent edges within the scene. The filter used for the FPF and the RIG is shown in Fig. 3.26. The RIG filter is a binary image of the target. The FPF is built from the same image with a fractional power of 0.5.

3.10 Object Isolation   57

Fig. 3.26. The filter

The images in Fig. 3.27 display the dynamic input for selected iterations. Iterations not shown did not produce interesting results (for some iterations the pulse field contains very few neurons firing). Some of the iterations shown did not alter the dynamic input, but the pulse outputs and correlation surface are interesting. They are presented to demonstrate the performance of the PCNN-FPF combination when non-target segments are being produced.

Table 3.2 contains the correlation peak and PCE for selected (all non-trivial iterations). The FPF was designed to produce a peak height of 1 as an impulse response. Acceptable correlation spikes have a peak greater than 0.25 and a PCE greater than 0.0010.

Table 3.2. Correlation response for nontrivial iterations

Iteration	Corr. Height	PCE
0	0.42	.0002
1	0.99	.0015
2	0.30	.0027
3	0.16	.0089
6	0.06	.0417
7	0.19	.0030
8	0.53	.0020
9	0.34	.0002
10	0.45	.0019
11	0.37	.0009
12	0.23	.0014
13	0.22	.0020
14	0.30	.0019

58   3 Automated Image Object Recognition

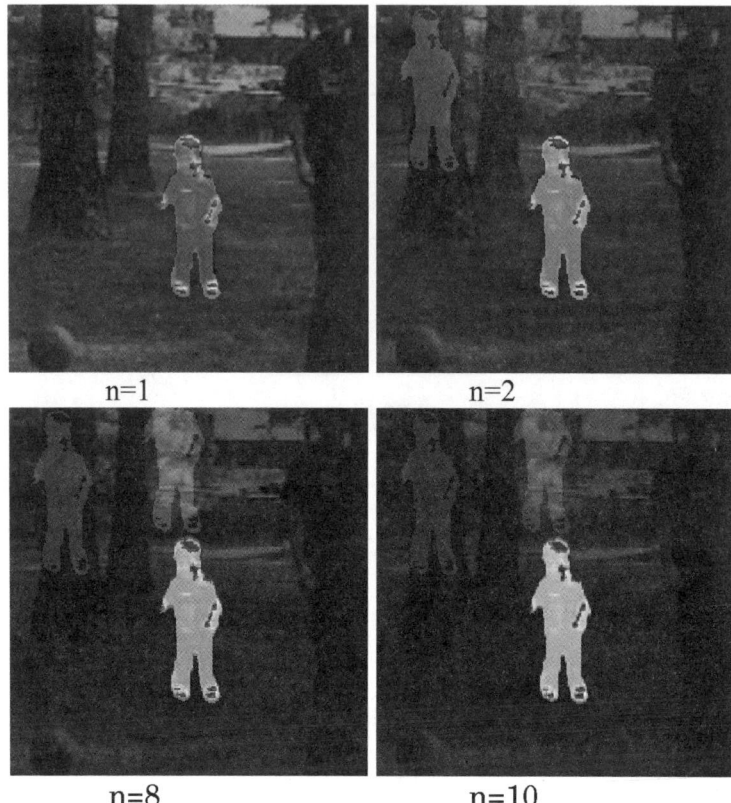

Fig. 3.27. The dynamic input in the feedback system

As is seen in Fig. 3.27 the target is gradually enhanced compared to the rest of the image. By the end of the demonstration the target is by far the most significant object in the scene. This is the desired result. The target object has been isolated.

3.11 Dynamic Object Isolation

Dynamic Object Isolation (DOI) is the ability to perform object isolation on a moving target. A system to do this has two alterations to the static object isolation system discussed above. The first is that it trains the filter on many differing views of the target and the second is that it must be optimised to recognise the target in frames that were not used in training. This second goal forces the system to exhibit generalisation abilities since the target may be presented differently in the non-training views. Using the example of the boy kicking the ball, the non-training views would show the boy in a different

Fig. 3.28. A sequence of five (**a**, **b**, **c**, **d** and **e**) input images

configuration (arms and legs and new angles), different orientations and scale (as he moves towards the camera).

The difference in the configuration of the system is that the filter is trained on several views of the boy. Consider the images in Fig. 3.28, which are a series of original images (a movie). Four of the inputs (Figs. 3.28a, b, c, and e) were used to create the FPF filter and feedback mask. Figure 3.28d. was not used in the training sequence but was reserved as a test image. Actually, several other frames (not shown) were also used as test images with similar results to Fig. 3.28d. The FPF filter is shown in Fig. 3.29. It is a composite image of

Fig. 3.29. The composite filter

60   3 Automated Image Object Recognition

Fig. 3.30. The progression of the dynamic input ($n = 3, 11, 18$ and 27, respectively)

four pulse images from each of the training images. The progression of the Dynamic Object Isolation (DOI) process for the image in Fig. 3.28d is shown in Fig. 3.30.

3.12 Shadowed Objects

The PCNN relies heavily upon the intensity of the input pixels. Thus, shadowed objects tend to produce a radically different response in the PCNN. A shadowed object is one that has a diminished intensity over part of itself. Consider an object that has two segments A and B, which originally have very similar intensities and therefore would pulse in the same iteration. A shadow falls upon segment B and its intensity is reduced. Now, due to its lower intensity, the segment pulses in frames subsequent to that of A. The object now has separate pulsing activity.

This can be a devastating effect for many PCNN-based architectures and it can destroy effective processing. In the object isolation architecture, however, the FPF has the ability to overcome the detrimental effects of shadows. Since the FPF has the fractional power set to include discriminatory frequencies and the pulse segments have sharp edges, a sufficient correlation occurs between the filter and the pulsing of only a part of the target. In other words,

3.12 Shadowed Objects   61

 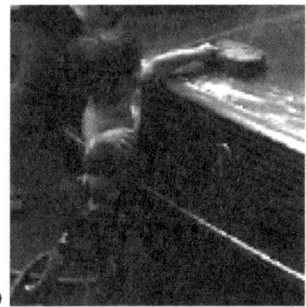

Fig. 3.31. a) The shadow mask. All pixels in the lower half of the image were used to decay the values of the original image, b) the shadowed input

Fig. 3.32. Progression of object isolation of a shadowed input

the filter can still provide a good enough correlation with segment A or B to progress the object isolation.

Consider the images in Fig. 3.31. In Fig. 3.31b is a shadowed image. This image was created from an original image in which the target (the boy) was cut-out and binarised (Fig. 3.31a). This binary image became the shadow mask and Fig. 3.31b was created by reducing all pixels in the lower half of the image that were ON in Fig. 3.31a. The effect is that the boy's shorts and legs were shadowed. The FPF filter and feedback mask were created with pulse images from the non-shadowed image.

The shadowed area intensity was sufficient to get the boy's shorts and legs to pulse in frames later than the torso and arms. However, the FPF filter was still able to find the partial target pulsing. The progression of the shadowed input is shown in Fig. 3.32.

Fig. 3.33. An input stimulus

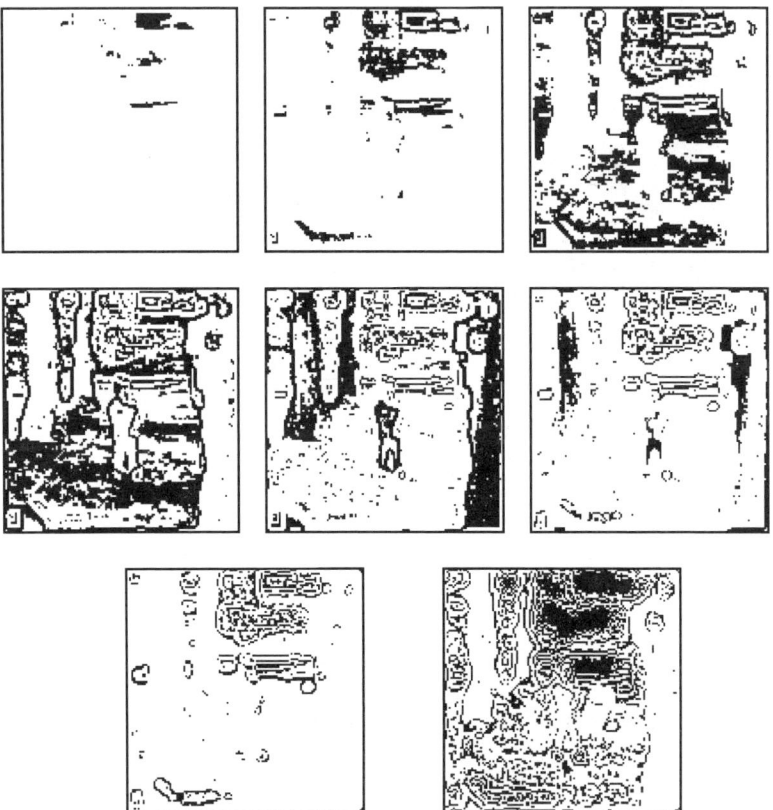

Fig. 3.34. Outputs of a PCNN with Fig. 6.1 as a stimulus

3.13 Consideration of Noisy Images

Random noise is an enemy of the PCNN. Pulse segments are easily destroyed by random noise. Noise can enter the system in three basic ways. The first is input noise in which S has noise added, the second is system noise in which noise is added to U, and the third is a random start in which Θ is initially

3.13 Consideration of Noisy Images   63

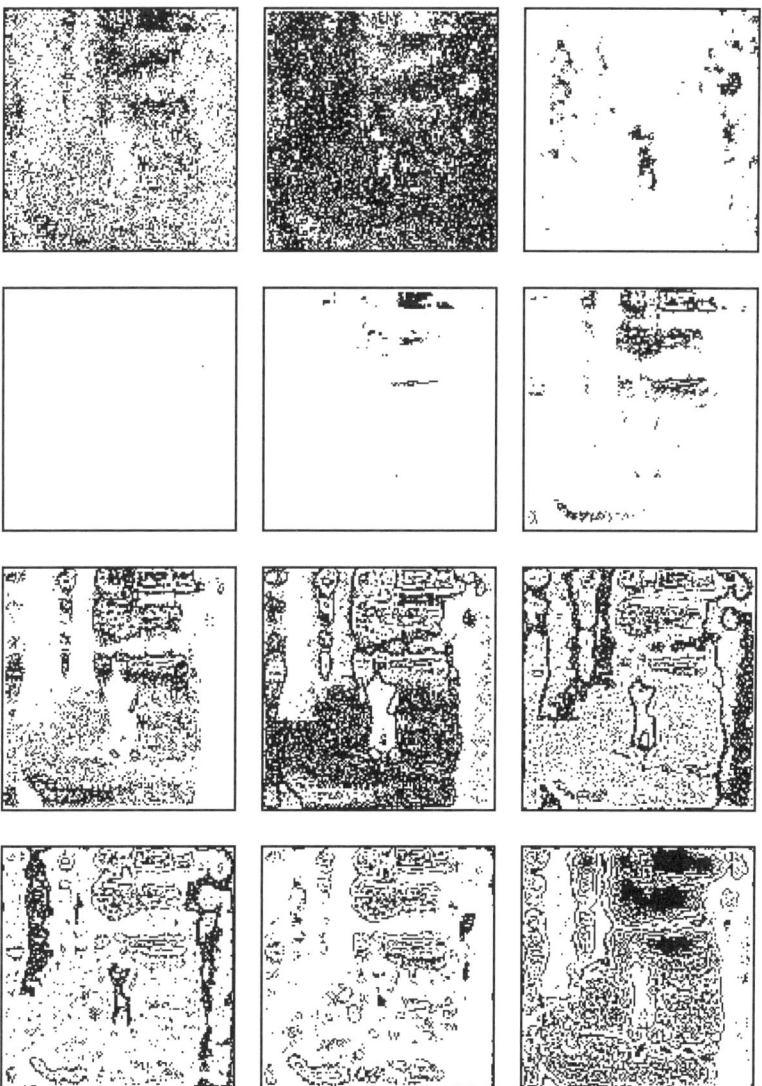

Fig. 3.35. Outputs of the PCNN with random initial threshold values

randomised. Any of these cases can destroy the PCNN's ability to segment. Consider the stimulus image shown in Fig. 3.33, which shows a boy (kicking a football), his father and some trees.

Using the input stimulus shown in Fig. 3.33, the original PCNN produces the temporal outputs of binary images shown in Fig. 3.34. Segmentation and edge enhancement are evident in the outputs shown. Compare these outputs to a system that initialised the threshold to random values between 0.0 and

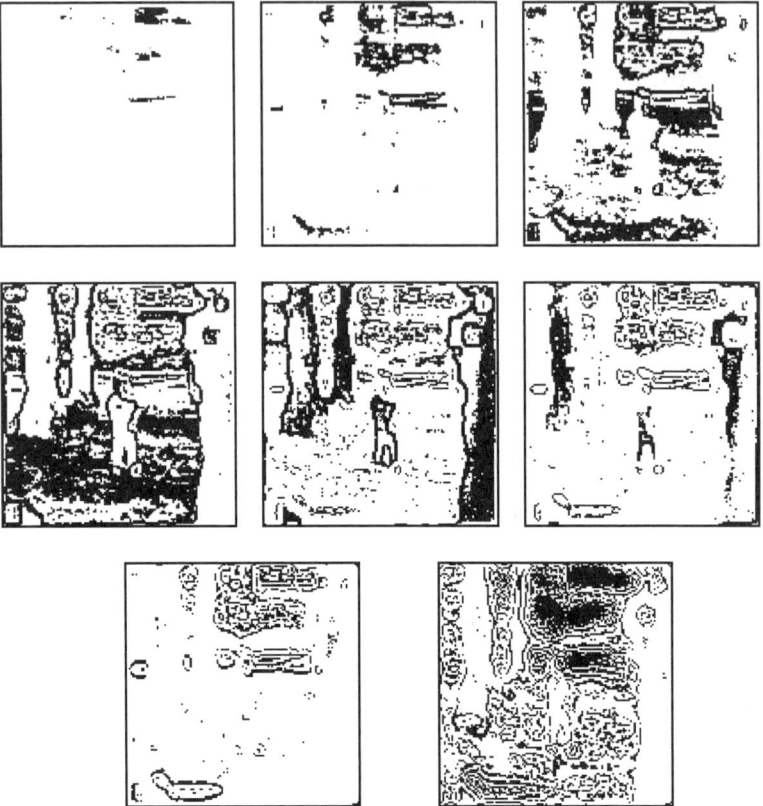

Fig. 3.36. Outputs of a PCNN with a signal generator

1.0. It should be noted that the initial values are less than 5% of threshold values after a neuron pulses. The results of this experiment are summarized in Fig. 3.35.

Certainly, the segments in the output are noisier than in the original case. This is expected. It should also be noted that the PCNN did not clean up the segments very well. There are methods by which this problem can be ameliorated.

The first method to be discussed for the reduction of noise uses a signal generator as a post-processor to the PCNN. This generator will produce small oscillations to the threshold values, which are in synchronisation with the natural pulsing frequency of stimulated neurons. The segments then tend to synchronise and noise is therefore significantly reduced.

A typical signal generator is the addition of a cosine term (where f is the design frequency) to the threshold during the creation of the output,

Fig. 3.37. Outputs of a PCNN with a signal generator and a noisy stimulus

$$Y_{ij}[n] = \begin{cases} 1 \text{ if } U_{ij}[n] > \Theta_{ij}[n-1] + (\cos(f*n/2\pi) + 1.0) \\ 0 \text{ Otherwise} \end{cases}, \qquad (3.13)$$

The outputs are shown in Fig. 3.36.

The noise of the system is now almost eliminated. The threshold levels have been synchronised by the continual addition and then subtraction of small values to the comparison function. The function periodically delays the pulsing of some neurons as the generator produces a larger output. When the generator produces a smaller output the pent-up potential of the neurons begins pulsing.

Noise can occur in other parts of the system. For example the input stimulus can be noisy. In Fig. 3.37 the output of a PCNN with a signal generator is shown for the case where the stimulus has random values between -0.1 and 0.1 added to the elements.

This system virtually eliminates the noise for the early iterations. However, the noise returns in subsequent iterations. The reason is quite simple.

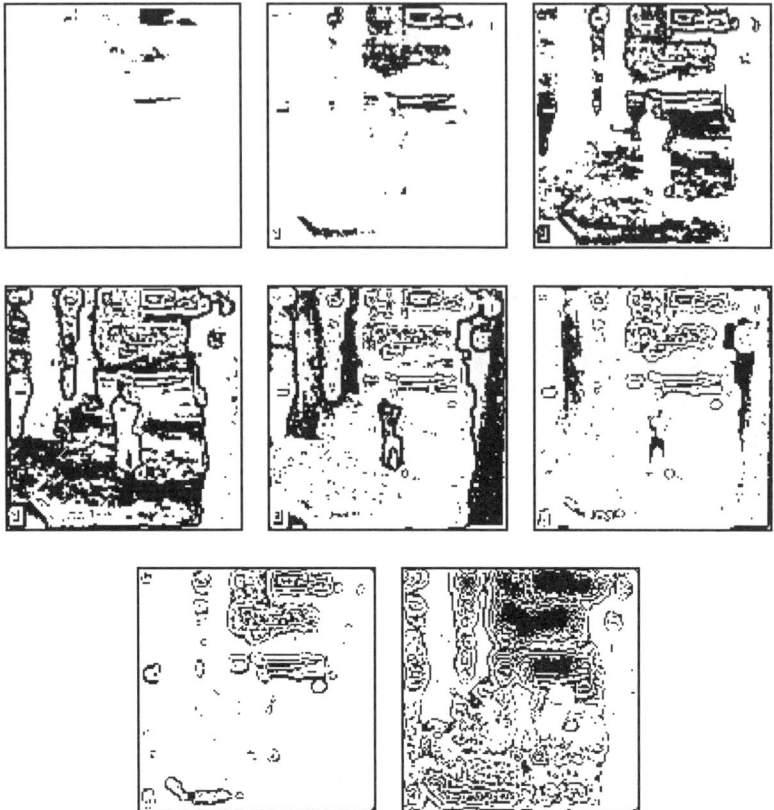

Fig. 3.38. Outputs of the PCNN after noise added to U for each iteration

Each iteration has the stimulus added to F. Thus, constant noise continually accumulates. Noise begins appearing in the output when the generator can no longer overcome the accumulation of the noise.

The last example adds dynamic noise to the system. In other words, our noise generator adds random zero mean noise is added to U each iteration. The values of the additional noise are $[-0.1, 0.1]$. The results of this test are shown in Fig. 3.38.

As can be seen, the noise is considerably reduced. In this case the noise was different for each iteration. These cancelling effects allowed the system to operate in a similar way to that of Fig. 3.37.

Another method of reducing the noise is to employ the fast linking algorithm. This was demonstrated earlier in Sect. 2.1.4 using the same example.

3.14 Summary

The PCNN and ICM are powerful engines in the fields of image processing and image recognition. The most important part of an image recognition algorithm is the extraction of information from an image. Without the proper information no recognition algorithm can perform.

The PCNN/ICM models have the ability to extract information that is pertinent to most applications. In this chapter the ability of these models to extract segments and edges was explored in several examples.

The ability to isolate segments as bright contiguous collections was demonstrated to be the foundation of several algorithms. Basically, one of the pulse iterations displays the target or a large portion of the target in an isolate manner allowing for easy recognition. Furthermore, this pulsation even occurs for dark targets. Thus, targets that are traditionally difficult for filters to find become easy to find in a pulse image.

4 Image Fusion

In a multi-spectral environment information about the presence of a target is manifest across the spectra. Detection of these targets requires the fusion of these different kinds of data. However, image fusion is difficult due to the large volume of data. Typically each detector channel does not provide enough information to detect the target with a significant level of confidence. Thus, each channel provides clues only and hints as to the presence of the target. Thus, it is practical to pursue methods that condense such a large volume of information to a more manageable size.

The Pulse-Coupled Neural Network (PCNN) has been shown to be a very powerful image processing tool [29,40,41,44] for single channel images. Its usefulness is its inherent ability to extract edges, texture, and segments from an image. The PCNN produces image autowaves that travel through the image. Autowaves are expanding waves that do not reflect or refract and annihilate each other as they collide. These autowaves are the key to the extraction of pertinent information from the image as will be demonstrated. The fusion process requires analysis of each channel of the image and the combination of these analyses. Thus, the image fusion process presented here will allow multiple PCNNs to create intra-channel autowaves and the novel addition of inter-channel autowaves. Thus, pertinent image formation is extracted from all channels in an interacting fashion. The solution to the image fusion problem proposed here uses a multi-channel PCNN and a FPF (fractional power filter) [26] to determine the presence of a target in a multi-channel image.

4.1 The Multi-spectral Model

The multi-spectral PCNN (εPCNN) is a set of parallel PCNNs each operating on a separate channel of the input with both inter- and intra-channel linking. This has a very interesting visual effect as autowaves in one channel cause the creation of autowaves in the other channels. The first autowave leads a progression of autowaves, but they all maintain the shape of the object. This is shown by the example in Figs. 4.1 and 4.2. The original image is a $256 \times 256 \times 24$, three channel (colour) image of a boy eating ice cream. The other images are the colour-coded pulse outputs of the three channels. Again the edges and segments are quite visible.

70   4 Image Fusion

Fig. 4.1. A three channel (colour) input image

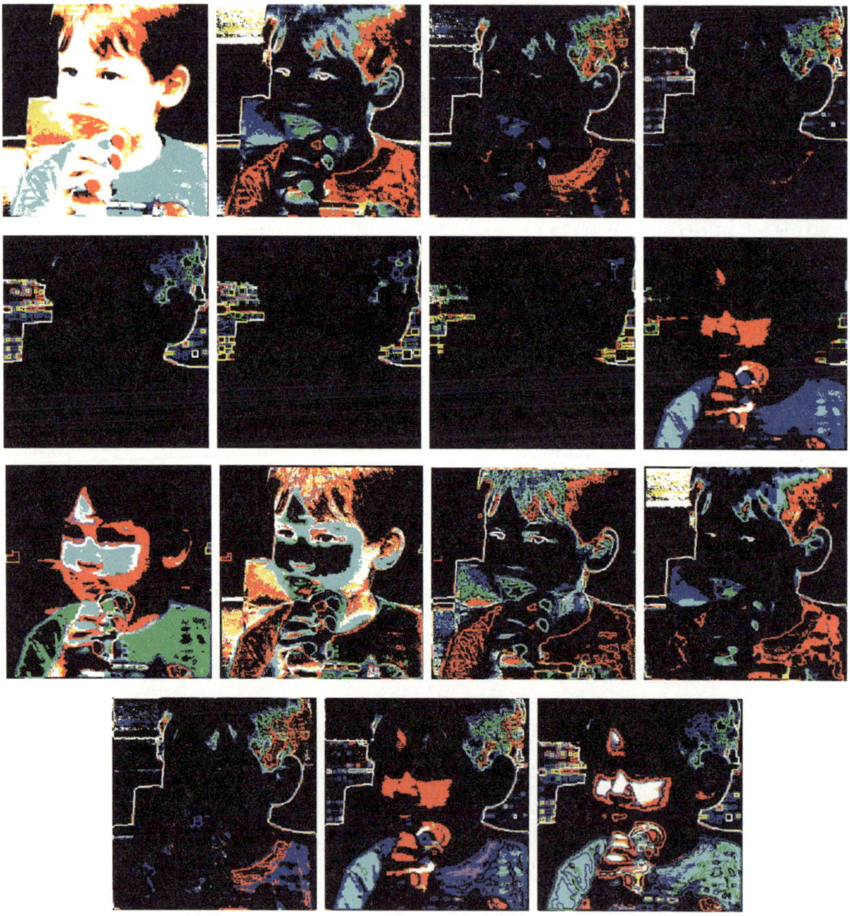

Fig. 4.2. The pulse outputs using the input shown in Fig. 4.1

An example of the cross channel linking is evident in the boy's hair. The hair is brown and changes in intensity. At $n = 1$ through $n = 3$ and $n = 8$ through $n = 14$ the autowaves are seen travelling down the boy's hair. The red autowave leads the others since the hair has more red than green or blue. The other autowaves follow, but all waves follow the texture and shape of the boy's hair.

The εPCNN equations require small alterations to account for the many channels denoted by ε,

$$F_{ij}^{\in}[n] = e^{\alpha_F \delta n} F_{ij}^{\in}[n-1] + S_{ij}^{\in} V_f \sum_{kl} M_{ijkl} Y_{kl}^{\in}[n-1], \tag{4.1}$$

$$L_{ij}^{\in}[n] = e^{\alpha_L \delta n} L_{ij}^{\varepsilon}[n-1] + V_L \sum_{kl} W_{ijkl}^{\varepsilon} Y_{kl}^{\varepsilon}[n-1], \tag{4.2}$$

$$U_{ij}^{\varepsilon}[n] = F_{ij}^{\varepsilon}[n]\left\{1 + \beta L_{ij}^{\varepsilon}[n]\right\}, \tag{4.3}$$

$$Y_{ij}^{\varepsilon}[n] = \begin{cases} 1 \text{ if } U_{ij}^{\varepsilon}[n] > \Theta_{ij}^{\varepsilon}[n] \\ 0 \text{ Otherwise} \end{cases}, \tag{4.4}$$

and

$$\Theta_{ij}^{\varepsilon}[n] = e^{\alpha_\Theta \delta n} \Theta_{ij}^{\varepsilon}[n-1] + V_\Theta Y_{ij}^{\varepsilon}[n]. \tag{4.5}$$

The intra-channel autowaves present an interesting approach to image fusion. The image processing steps and the fusion process are highly intertwined. Many traditional image fusion systems fuse the information before or after the image processing steps, whereas this system fuses during the image processing step. Furthermore, this fusion is not statistical. It is syntactical in that the autowaves, which contain descriptions of the image objects, are the portions of the image that cross channels. This method is significantly different from tradition in that it provides higher-order syntactical fusion.

4.2 Pulse-Coupled Image Fusion Design

There are several reasons to fuse the inputs of images of the same scene. One example is to fuse an infra red image with a visible image to see relations between items only seen in one of the inputs. The same detector image may also be filtered differently in order to enhance features of different, but related origin. Generally one fuses the signals from several sensors to get a better result. The system shown in Fig. 4.3 uses a multi-channel PCNN (εPCNN) *and* FPF to fuse the information in many images to reach a single decision as to the presence of a specified target and its location. The εPCNN creates inter- and intra-channel linking waves and multi-channel pulse images [42,43].

The advantages of the εPCNN include the inherent ability to segment an image. It does this without training or any knowledge of the target, and,

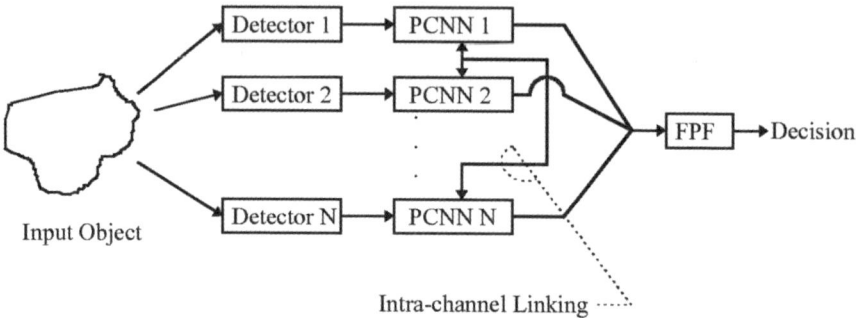

Fig. 4.3. Logic schematic of the multi-channel PCNN

therefore, it cannot identify a target. The FPF is an identification tool that allows for training of multiple targets or views. However, it is like all Fourier filters in that it performs on general, 'real-world' inputs. It performs well on clean, binary inputs, which is exactly what the εPCNN produces. The logical course of action, then, is to use the segmentation ability of the εPCNN to provide inputs to the FPF.

Each output pulse wave is a real binary image, and the input to the FPF is a phase-encoded accumulation of the pulse images. The phase encoding process basically allows for the combined representation of several channels in a single complex array. This is possible because the pulse images are binary in each channel and cross talk is minimal. Phase encoding of real arrays (non-binary) would create a significant amount of cross talk and would usually destroy a significant amount of information. However, binary images due to their extremely limited dynamic range can create unique phase encoded images. In many instances the original binary images can be re-created from the phase-encoded image, thus indicating that the encoding process does not destroy information. In the case of three channels, the elements of the phase-encoded image can only realise one of eight different values which are uniquely created by the eight possible binary value combinations.

Thus, the final output of the multi-channel εPCNN is a complex image that contains data from all channels,

$$Y^T = \sum_\varepsilon Y^\varepsilon e^{i2\pi\varepsilon/N} . \tag{4.6}$$

The FPF is trained on a multi-channel target selected in the original images. For training views of the target are cut-outs from a few of the original images. Each training image is

$$x_i^\varepsilon = e^{i2\pi\varepsilon/N} S^\varepsilon \tag{4.7}$$

and trained with the corresponding constraint

$$c_i = e^{i2\pi\varepsilon/N} \tag{4.8}$$

and the filter is trained according to (3.6)–(3.8).

The final result is the correlation, Z,

$$Z = h \otimes Y^T. \tag{4.9}$$

The presence of a target will produce a large correlation signal at the location of the target. An exact match will produce a signal of height N. Stimuli that do not exactly match the training targets will have a lower correlation value depending upon the fractional power.

The data set and the defined task assist in determining the value of p. If the data set varies greatly then p can be lowered. If the targets consist mainly of low frequency data, the mid-point of the trade-off will move to a higher value of p. If the task requires a higher precision of discrimination the user should give a larger value of p. Generally, the best way to determine the proper value of p is to try several values. The entire fusion algorithm follows the following prescription:

i. Given a defined target training set, X, the FPF filter is generated.
ii. Given a stimulus, S, with ε channels each channel is separately represented by S^ε. All arrays of the PCNNs are initialised to 0.
iii. One εPCNN iteration is performed following (4.1)–(4.5).
iv. The outputs Y^e are phase-encoded to form Y^T as by (4.6).
v. The correlation Z is calculated to identify the presence of a target by (4.9). A large correlation spike is indicative of the presence of a target.
vi. Steps 3–5 are repeated until a target is clearly evident or several iterations have produced no possible target identifications. The actual number of iterations is problem dependent and currently determined by trial and error. Generally, 10–20 iterations are sufficient.

4.3 A Colour Image Example

The example below uses the colour-input image in Fig. 4.1. This image is a three-channel image. The εPCNN responses are also shown.

The target was selected to be the boy's ice cream. Cutting out and centring the ice cream created the FPF training image. In traditional filtering, creating a training image from a cut-out may be dangerous. This is because the cutting process induces artificial edges. These can constitute Fourier components. These generally differ significantly from the original image and may consequently change the correlation result. In other words, the filter may have large Fourier components that will not exist in the input. The PCNN, however, produces pulse images, which also contain very sharp edges, so in this case cutting out a target is not detrimental. Both components of the correlation contain sharp edges so the cutting-out process will not adversely affect the signal to noise of the correlation.

The FPF with $p = 0.8$ is shown in Fig. 4.4. Each of the εPCNN pulses was correlated with the filter. The correlation surfaces for the first four nontrivial

Fig. 4.4. The complex Fractional Power Filter (FPF) of the target

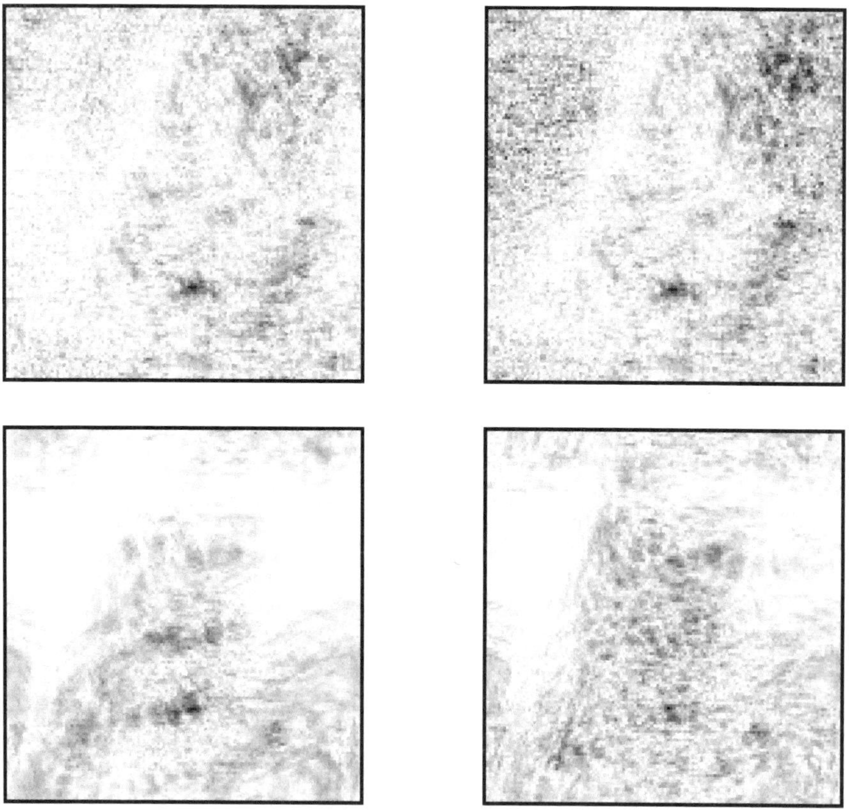

Fig. 4.5. Correlation surfaces of the first four non-trivial PCNN responses. Images shown are for $n = 1, 2, 7$ and 8

multi-channel pulses are shown in Fig. 4.5 (where a trivial output is one in which very few neurons pulse).

The trivial and some of the non-trivial output pulse images do not contain the target. In these cases no significant correlation spike is generated. The presence or absence of a target is information collected from the correlation of several pulse images with the filter. The target will produce a spike in several of these correlation surfaces. The advantage is that the system tolerates a few false positives and false negatives. An occasional correlation spike can be disregarded if only one output frame in many produced this spike, and a low target correlation spike from one pulse image can be overlooked when many other pulse images produced significant spikes at the target location. This method of the accumulation of evidence has already been shown in [40].

The correlation surface for all shown iterations display a large signal that indicates the presence of the target. It should be noted that the target appears only partially in any channel. From the multi-channel input a single decision to indicate the presence of the target is reached. Thus, image fusion has been achieved.

As for computational speed, the PCNN is actually very fast. It contains only local connections. For this application $M = W$ so the costly computations were performed once. For some applications it has been noted that using $M = 0$ decreases object crosstalk and also provides the same computational efficiency. Also quick analysis of the content of Y can add to the efficiency for the cases of a sparse Y. In software simulations, the cost of using the FPF is significantly greater than using the εPCNN.

4.4 Example of Fusing Wavelet Filtered Images

In this example we will compare the results of a single PCNN to a simple fused PCNN. The single PCNN will receive only one input, the grey scale image of 'Donna'. The fused PCNN will send this input to the central PCNN, and also to two wavelet filtered inputs to two adjacent PCNNs, which are fused with the central PCNN. The results are compared in Figs. 4.6 and 4.14.

4.5 Detection of Multi-spectral Targets

This demonstration uses images collected by an experimental AOTF (acousto-optical tunable sensor) [38, 39] An AOTF acts as an electronically tunable spectral bandpass filter. It consists of a crystal in which radio frequencies (RF) are used to impose traveling acoustic waves in the crystal with resultant index of refraction variations. Diffraction resulting from these variations cause a single narrowband component to be selected from the incoming broadband scene. The wavelength selected is a function of the frequency of the RF signal applied to the crystal. Multiple narrowband scenes are be generated

Fig. 4.6. The original input (**top**) and two wavelet filtered images (here shown inverted) are the three inputs to the fused PCNN

Fig. 4.7. 4102dt0 (long wave) and 4130dt0 (short wave)

by incrementally changing the RF signal applied to the transducer. The selected wavelength is independent of device geometry. Relative bandwidth for each component wavelength is set by the construction of the AOTF and the crystal properties. The sensor is designed to provide 30 narrow band spectral images within the overall bandpass of 0.48–0.76 microns.

4.5 Detection of Multi-spectral Targets 77

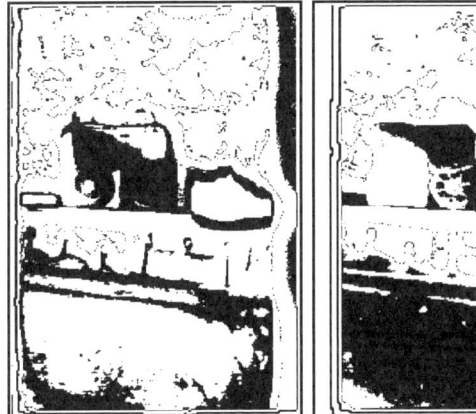

Fig. 4.8. The output for $n = 7$ and channels 18 and 2

Figure 4.7 contains examples of individual channel input images. The image designated as 4102dt0 is near the long wavelength end of the spectrum. The second example, 4130dt0 is the shortest wavelength image. Figure 4.8 contains examples of individual channel binary outputs of the PCNN. These specific examples were chosen because they display features associated with the mines. Figure 4.9 displays a gray scale representation of all channels in particular iterations. The gray encoding coarsely displays phase. As can be seen the targets become quite visible in detail for some iterations (e.g. $n = 7$).

Figure 4.10 displays the amplitude of the spiral filter built to detect one of the targets by the FPF method. Figure 4.11 displays a 3-D plot of the correlation surface between the filter and iteration 7.

Figure 4.12 displays a cross-sectional slice of the correlation surface that passes through the peak of the correlation (labeled surf07b). As can be seen, the correlation produces a signal that is significantly greater than the noise. Similar performance was obtained from the filter built for the other target. The other plot in the figure is the correlation between the spoked landmine (immediately attenuated by a Kaiser window) and the original image of channel 20. As can be seen that this correlation function does little to indicate the presence of a target. The large signal is from the Halon plate. A drastic improvement is seen between this correlation and that produced through the spiral filter.

Not all iterations will contain the target as a segment. This is the inherent nature of the pulse images. The particular iteration in which the target appears is dependent upon scene intensity. It may coincidental that immediately neighboring objects may pulse in the same iteration as the target making it difficult to distinguish the border between the two objects in the output. FPF correlations may still produce a significant correlation if a majority of targets edge is present and other iterations will separate these neighboring objects.

Fig. 4.9. Gray scale representations of the multi-channel pulse images

4.5 Detection of Multi-spectral Targets   79

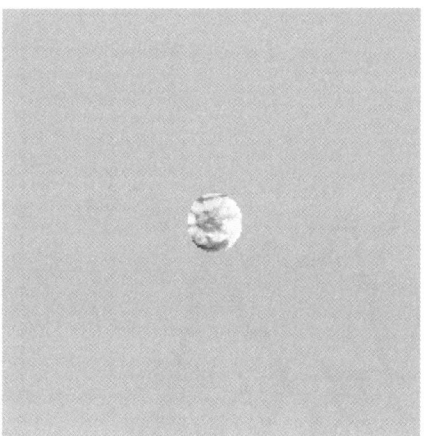

Fig. 4.10. The amplitude of a spiral filter

Fig. 4.11. The correlation of the filter with a portion of the iteration 7. Dark pixels indicate a higher response

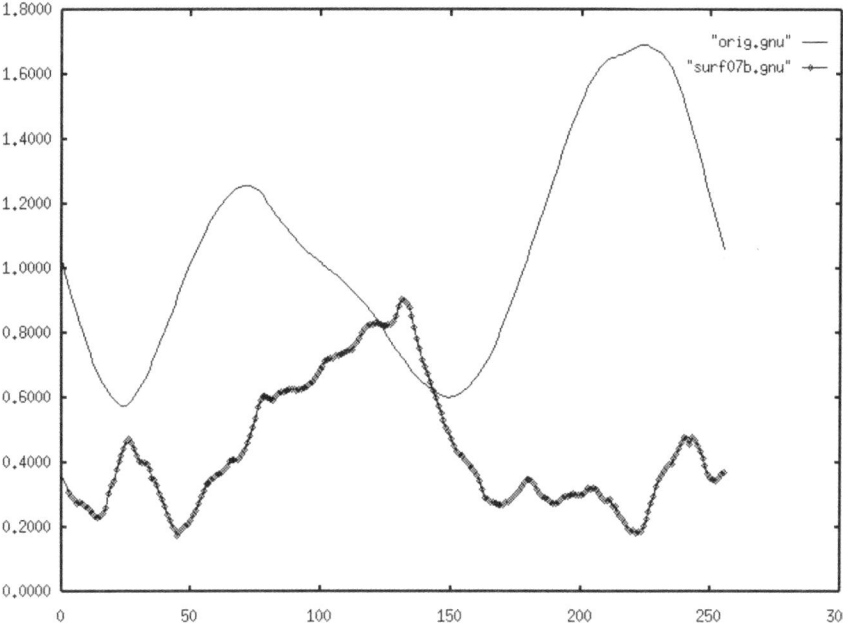

Fig. 4.12. Cross-sections of a correlation surface for the target and the original image (orig.gnu) and the spiral filter with pulse image $n = 7$

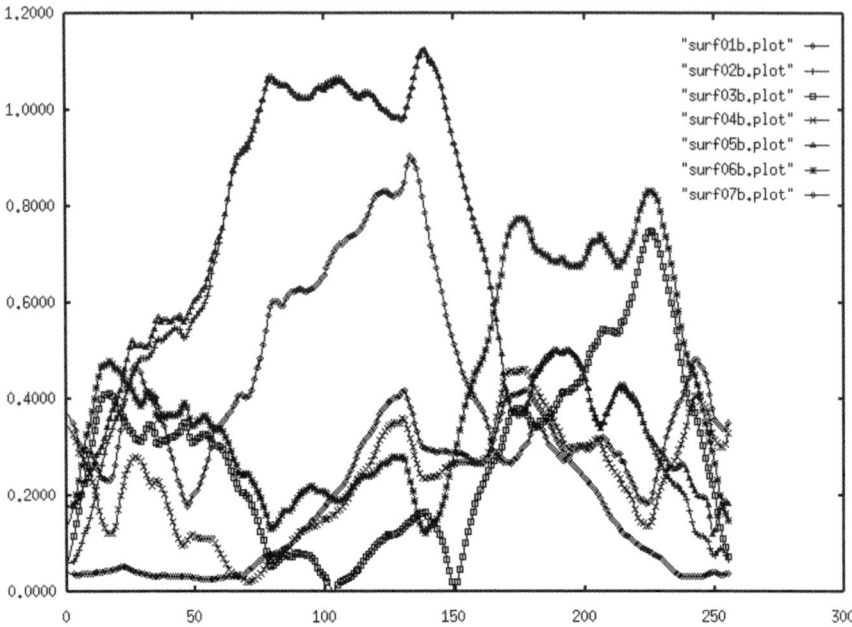

Fig. 4.13. Cross-sections of the spiral filter with coded pulse images from several iterations

Figure 4.13 displays a 1D correlation slices (through the target region) for the first seven iterations. Iterations $n = 2$ and $n = 5$ produced the largest correlation signals, but the width of the signals (caused by the neighboring location of the two landmines and their similarity in overall shape and size) prevents these iterations from being indicative of the target landmine. The Halon plate also causes significant signals due to its very high intensity which can be highly reduced by normalizing the correlation surface with a smoothed input intensity image. Iteration $n = 7$ (which is also displayed in Fig. 4.12) indicates that a target exists.

4.6 Example of Fusing Wavelet Filtered Images

In this example we will compare the results of a single PCNN to a simple fused PCNN. The single PCNN will receive only one input, the grey scale image of 'Donna'. The fused PCNN will send this input to the central PCNN, and also to two wavelet filtered inputs to two adjacent PCNNs, which are fused with the central PCNN. The results are compared in Fig. 4.14.

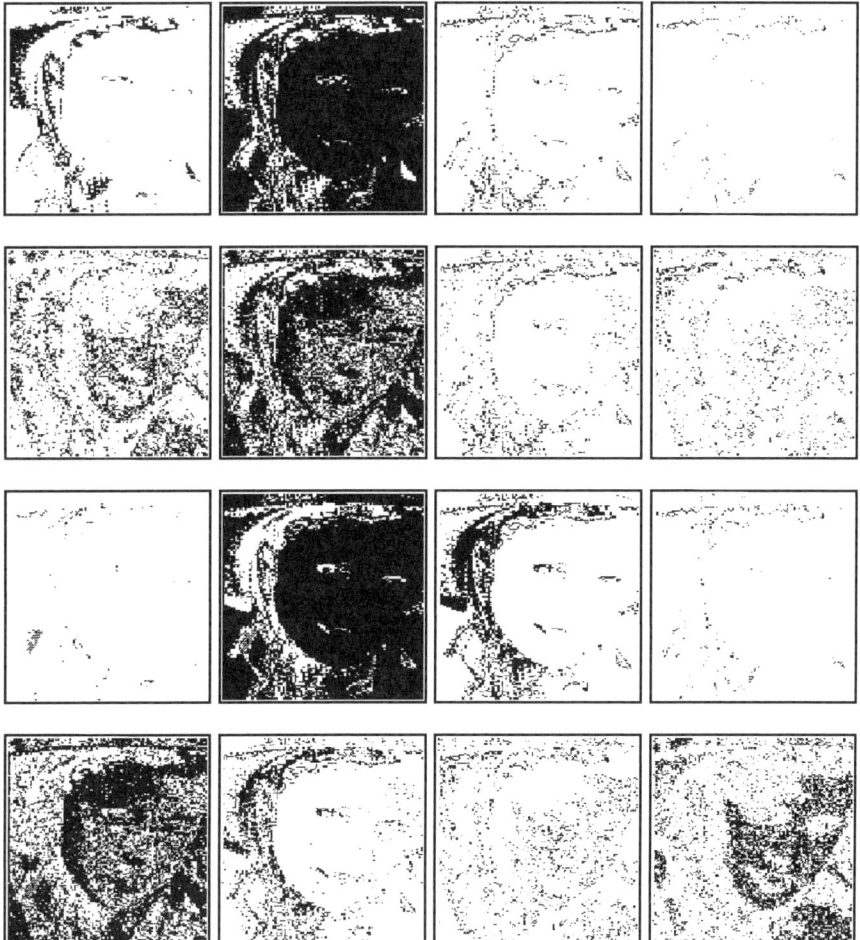

Fig. 4.14. Eight outputs from a fused PCNN (**top** two rows) and a fused PCNN (**bottom** rows)

4.7 Summary

In this chapter we introduced a multi-channel εPCNN and coupled it with an FPF with complex constraints. Image fusion is achieved by allowing the εPCNN to create pulse images for each channel that are coupled with the other channels. Analysis of the pulse images can then be performed by a complex FPF. The phase degree of freedom allows for the encoding of multiple channels in a single filter. The εPCNN is used to extract both spectral and intensity features, which can be easily analysed by this filter.

5 Image Texture Processing

In many applications the information that is important is the textures within an image. There are many such applications and in this chapter the use of texture analysis using medical images will be considered.

Regions in an image pulse in unison when their stimuli are the same. If the stimuli are varied (there exists a texture) then the synchronised behaviour of the pulse segments will disintegrate. The regions become more desynchronised as the cycles progress. This desynchronisation is dependent upon the texture of the input and thus texture can be measured and used for segment classification.

The authors would like to acknowledge the significant contribution of Guisong Wang for the material in this chapter.

5.1 Pulse Spectra

Consider again the images shown in Fig. 3.6 and 3.7. The nucleus of the red blood cell contains a texture. In iteration $n = 1$ the nucleus pulses as a segment completing the first cycle. The second cycle occurs in iterations $n = 16$–19. The neurons of the nucleus have desynchronised and the pulses are separated according to the texture of the original segment. Thus, measurement of texture is performed over several iterations rather than in a single iteration.

Texture is an interesting metric in that it describes a property that spans several pixels but in that region those pixels differ. It is a segment described by dissimilarity. The size of this region, however, is defined by the user and cannot be set to a uniform distance.

There have been several methods by which texture has been measured. Many of these rely on statistical measures but the ICM is different. The higher order system can also extract relational information.

The image in Fig. 5.1 displays two different cloth textures [54]. Even though the pixel values vary on a local scale the texture is constant on a global scale.

One of the simplest methods of measuring texture is to simply measure the statistics such as the mean, the variance (and the coefficient of variance), skewness, and kurtosis. The mean for a vector of data is defined as,

84   5 Image Texture Processing

Fig. 5.1. Example textures

$$m = \frac{1}{N}\sum_{i=1}^{N} x_i, \tag{5.1}$$

where V and H are the vertical and horizontal dimensions of the image. The variance and the coefficient of variance are defined as,

$$s = \frac{N\sum_{i=1}^{N} x_i^2 - m^2}{n(n-1)}, \tag{5.2}$$

and

$$cv = s/m. \tag{5.3}$$

The skewness and kurtosis are higher order measures and are defined as,

$$\tau = \frac{N}{(N-1)(N-2)} \sum_{i=1}^{N} \left(\frac{x_i - m}{s}\right)^3, \tag{5.4}$$

and

$$k = \frac{N(N+1)}{(N-1)(N-2)(N-3)} \sum_{i=1}^{N} \left(\frac{x_i - m}{s}\right)^4 - \frac{3(N-1)^2}{(N-2)(N-3)}. \tag{5.5}$$

For simple images like Fig. 5.1 it is possible to measure and distinguish the textures according to these measures. The values for the two sample images are shown in Table 5.1.

Table 5.1. First order texture measures

	Texture 1	Texture 2
Mean	0.492	0.596
Variance	0.299	0.385
Coefficient of var.	0.607	0.647
Skewness	0.032	−0.058
Kurtosis	0.783	0.142

5.1 Pulse Spectra 85

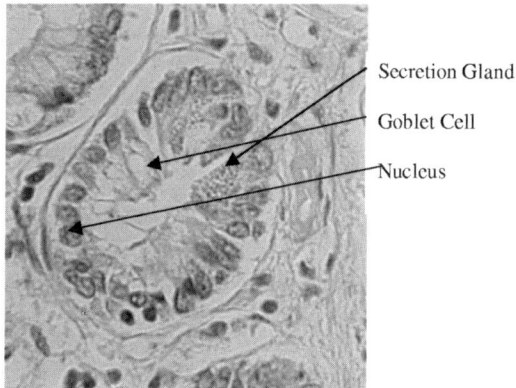

Fig. 5.2. An image

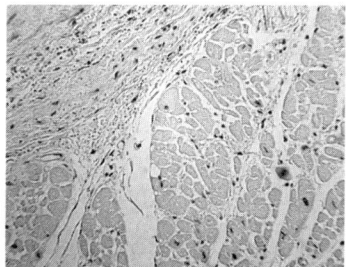

Fig. 5.3. An image

However, real problems generally do not fill the image frame with a single texture. Quite often it is desired to segment the image according to the texture implying that the texture boundaries are not known but rather need to be determined.

The image in Fig. 5.2 displays a secretion cell with a variety of textures. One typical application would be to segment this image according to the inherent textures.

To employ the ICM to extract textures data is extracted for each pixel for all iterations. Since texture is defined by a region of pixels rather than a single pixel the pulse images are smoothed before the texture information is extracted. The information taken from a location (i,j) defines the pulse spectrum for that location and is defined by,

$$p_{i,j}[n] = M\{Y\}_{i,j}[n]. \qquad (5.6)$$

Here the function $M\{\}$ is a smoothing operator over each pulse image. The goal is that all pulse spectra within a certain texture range will be similar. The images in Figs. 5.3 and Fig. 5.4 display an original image and a few of the pulse images. Common textures pulse in similar iterations.

86   5 Image Texture Processing

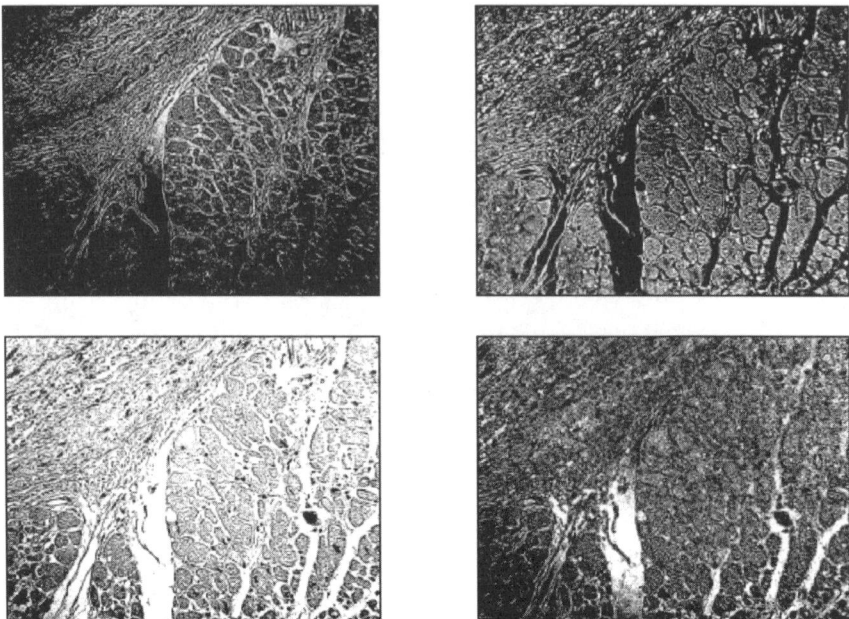

Fig. 5.4. Pulse images for $n=1$, $n=2$, $n=8$, and $n=9$

This displays only four of many pulse images. In this case the number of iterations was selected to be twenty.

For the case of standard textures (similar to Fig. 5.1) the method of using pulse spectra was compared to other methods. These methods are listed with their citations but their methods will not be reviewed here.

- Autocorrelation (ACF) [53, 56]
- Co-occurrence matrices (CM) [47, 48]
- Edge frequency (EF) [53, 56]
- Law's masks (LM) [51]
- Run Length (RL) [53, 56]
- Binary stack method (BSM) [45, 46]
- Texture Operators (TO) [57]
- Texture Spectrum (TS) [49]

In [55] the performance of all of the methods in the above list were compared on a standardized database of textures. The tests consisted of training on all but one of the images and then that image was used as a test. This test was repeated several times using each of the images as the one not used in training. Recall used the k-nearest neighbors algorithm and the results are shown in Table 5.2 for different values of k. At the top of this chart the texture recognition method using the ICM was added and it can be seen that it rivals the best performing algorithms.

Table 5.2. Recognition rates for various texture models

Texture Analysis Methods	$K=1$	$K=3$	$K=5$	$K=7$	$K=9$
ICM	94.8%	94.2%	93.9%	92.1%	91%
ACF	79.3%	78.2%	77.4%	77.5%	78.8%
CM	83.5%	84.1%	83.8%	82.9%	81.3%
EF	69%	69%	69.3%	69.7%	71.3%
LM	63.3%	67.8%	69.9%	70.9%	69.8%
RL	45.3%	46.1%	46.5%	51.1%	51.9%
BSM	92.9%	93.1%	93%	91.9%	91.2%
TO	94.6%	93.6%	94.1%	93.6%	94%
TS	68.3%	67.3%	67.9%	68.5%	68.1%

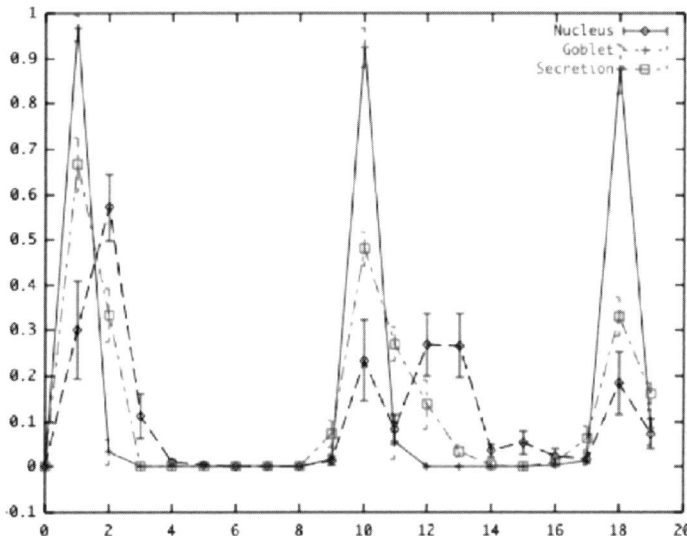

Fig. 5.5. Plots of the average and standard deviation of the three selected regions of Fig. 5.2

5.2 Statistical Separation of the Spectra

The real task at hand is to measure the textures of a complicated image as in Fig. 5.2. A requirement for the accomplishment of this task is that the spectra discriminate between the different textures. This means that the spectra in one texture region must be similar and compared to another region they must be dissimilar.

To demonstrate this three regions in Fig. 5.2 were selected for testing. This image is 700×700 pixels and each region selected was only 10×10 at the locations marked in the figure. The average and standard deviation of the spectra for each region are shown in Fig. 5.5.

The desire is to have each average signature differ significantly from the others and to also have small standard deviations. Basically, the errorbars should not overlap. Clearly, this is the case and therefore discrimination of texture using the ICM is possible.

5.3 Recognition Using Statistical Methods

A simple method of classifying regions in an image by the texture is to simply compare a pulse spectrum to all of the average spectra in a library. The library consists of average spectra from specified training regions (as in Fig. 5.5). This is similar to the procedures practiced in multi-spectral image recognition.

For each pixel in an image there is a pulse spectrum and this can be compared to those in a library. The pixel is then classified according to which member of the library is most similar to the pixel's spectrum. A pixel's spectrum can be classified as unknown if it is not similar to any of the members of the library. For this example, the elements of the spectrum needed to be within one standard deviation of the library spectrum in ordered to be considered close. This measure exclude spectrum members that were close to 0. More formally, the spectrum of the pixel in the image is defined as d_i where $i = 1, 2, \ldots, 20$ (the number of iterations in the ICM). The library consists of a set average spectra, m_i^k, where k is the index over the number of vectors in the library. For each member of the library there is also the standard deviation of the elements, σ_i^k. The pulse spectrum is considered close if for all $d_i > \varepsilon$,

$$\left| d_i - m_i^k \right| < \sigma_i^k, \tag{5.7}$$

where ε is a small value greater than 0.

Using this measure the pixels in the image of Fig. 5.2 that were classified as belonging to the *nucleus* class are shown (black) in Fig. 5.6a. In Fig. 5.6b the pixels classified as *secretion* are shown and in Fig. 5.6c the pixels classified as *goblet* are shown.

In this example many of the pixels were classified correctly. However, there were a lot of false positives. Part of the problem is that the texture of some of these mis-classified regions is very similar to that of the target. For example there are many nuclei outside of the large cell that have similar texture to the nuclei inside of the cell. Likewise, the goblet cells have similar texture to many regions outside of the cell. These strong similarities make it a difficult problem to solve.

Another cause of these false positives is that the texture of regions of similar class are somewhat different. The texture of the individual nuclei inside of the large cell are different.

Unfortunately, it is quite common to have a problem in which the texture of target regions different more than the texture between non-target and

5.4 Recognition of the Pulse Spectra via an Associative Memory

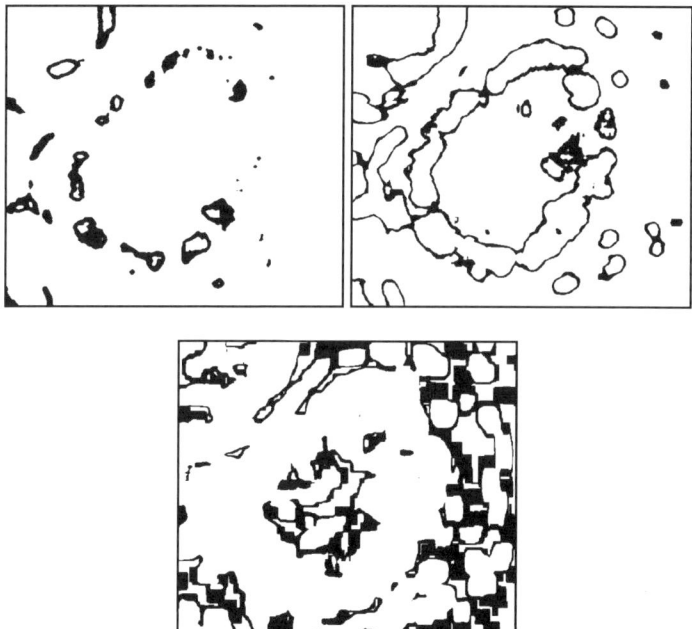

Fig. 5.6. The classification by texture of the pixels as *nucleus*, *secretion*, and *goblet*

target regions. If this wasn't the case then this would be an easy problem to solve.

In this situation the classification system is insufficient. In other words, the use of statistical comparison between the spectra is incapable of discriminating between target spectra and non-target spectra. A much more powerful discrimination algorithm is required.

5.4 Recognition of the Pulse Spectra via an Associative Memory

The inability of the previous system to completely recognize a texture may be caused by either of two problems. The first may be that the texture extraction engine is inadequate. The second is that the process that converts extracted texture to a decision may be inadequate. In the previous case the decision was reached by simply comparing statistics of the texture. However, in a case in which one class may contain more than one texture this decision process will be inadequate. Thus, we attack the problem with a stronger decision making engine.

There are several types of associative memories that are available to be used and certainly a system that contends optimality would consider several of these memories. In the case here, the goal is to demonstrate that the ICM can extract sufficient information from an image, thus only an adequate associative memory need be considered. If the combination of the ICM and the chosen associative memory sufficiently classify the image then the contention is that the ICM sufficiently extracted the texture information from the image.

The associative memory used here is a simple greedy algorithm that is fast and effective [50]. The philosophy is that simple decision surfaces are created and deleted as necessary. They are, however, never moved. This is a different philosophy than a neural network which establishes a set number of decision surfaces by defining the number of hidden neurons at the onset of training. Then the training process moves these decision surfaces to optimize the recall of the training data. In the system used here the decision surfaces are simply created and destroyed.

Consider a set of training data represented by a set of D-dimensional input vectors $\boldsymbol{x}_n$ that are each associated with a binary scalar output y_n. The goal is to create an associative memory such that the output of the system y'_n is sufficiently close to the desired output y_n for all n, and $y'_n = F\{\boldsymbol{x}_n\}$.

The process begins by iteratively considering each association. The output is binary and therefore can have one of two states, V and W. For the sake of argument, let's assume that the first training pair is $\boldsymbol{x}_1 : V$. So, the point in D-space defined by $\boldsymbol{x}_1$ is defined to have a value V. At this time no other training pairs have been considered and therefore we can declare all points in D-space to have a value V. This can be done since information about the existence of any other value is absent. If the next training pair to be considered is $\boldsymbol{x}_2 : V$ then our contention is not violated and therefore no training need to take place.

However, if the second training pair in the data set it $\boldsymbol{x}_2 : W$ then there is a violation. Now, there are two points in D-space which have different values. Absent any other *a priori* information the space between these two points is divided by a decision plane. Any point on the $\boldsymbol{x}_1$ side of the decision plane is declared as V and any point on the other side is declared a W. Thus, a decision surface was added.

Each training pair in the data set is considered and if it violates the current state of the system then decision surfaces between it and other training points are added as necessary. The addition of decision surfaces may supersede previous decision surfaces. For example, consider a cased in which one decision surface divides $\boldsymbol{x}_1$ and $\boldsymbol{x}_2$ and a second decision surface is later added to divide $\boldsymbol{x}_1$ and $\boldsymbol{x}_3$. However, this second surface also divides $\boldsymbol{x}_1$ and $\boldsymbol{x}_2$. Therefore, the first decision surface is no longer needed at it can be removed.

The process continues until each of the training data pairs has been considered. When the process is complete the system will be able to accurately recall each of the training data pairs. The recall system simply considers and input vector $\boldsymbol{x}$ and determines which side of each decision surface it is on. This information is then compared to that of each of the training vectors

5.4 Recognition of the Pulse Spectra via an Associative Memory

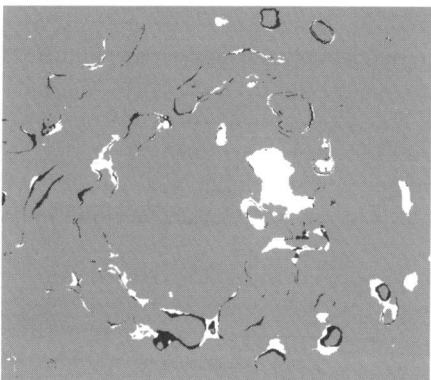

Fig. 5.7. Classification of the pixels to the secretion class

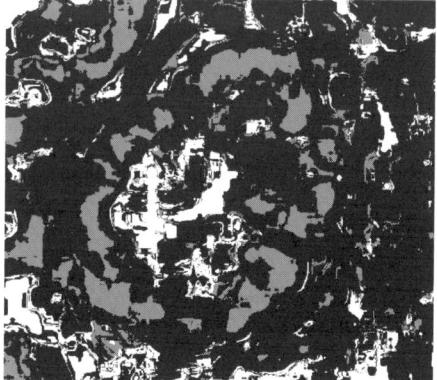

Fig. 5.8. Classification of the pixels to the goblet class

and the input is classified according to the training vector that has similar decisions.

For the texture application the input vectors are the pulse spectra and the outputs is a declaration as to whether a spectra belongs to a class or not. It is a binary decision. Thus, if there exist N classes then there will need to be N associative memories as each one only has the ability to declare whether a point is in a class or not.

The pixels used in training in the statistical example were also used here. Figure 5.7 displays that classification of pixels in the *secretion* class. All of the white pixels are declared to be in the class, all of the gray pixels are declared to be out of the class, and all of the black pixels are undefined. In this case the input vector to the associative memory produced a set of decisions that were not similar enough to any of the training vectors.

Figure 5.8 contains the classification of the pixels for the *goblet* class. In this case many of the pixels are classified as not known, however, the goblet pixels are correctly classified.

Again the intent was to demonstrate that the ICM has the ability to extract texture information from an image. It was not to build the perfect texture recognition engine. Clearly, large portions of the above cases are classified correctly, thus indicating that the texture information.

5.5 Summary

The ICM will create a set of pulse images from an input image and these pulse patterns are dependent upon the image texture. Thus, it is possible to extract texture information from the pulse images. This is accomplished by extracting a pulse spectrum for each pixel in the image. Similar to the methods used in the multi-spectral image recognition systems, these spectra are used to classify the pixels according to their texture.

6 Image Signatures

With the advent of the cheap digital camera we have the ability to overwhelm ourselves with digital images. Thus, there is a need to be able to describe the contents of images in a condensed manner. This description must contain information about the content of the images rather than just a statistical description of the pixels.

Measurements of the activity in the brain of small mammals indicate that image information is converted to small one-dimensional signals. These signals are dependent upon the shapes contained in the input stimulus. This is a drastic reduction in the amount of information used to represent the input and therefore is much easier to process.

The goal is then to create a digital system that condenses image information into a signature. This signature must be dependent upon the contents of the image. Thus, two similar signatures would indicate that the two images had similar content. Once this is accomplished it would be fairly easy to construct an engine that could quickly find image signatures similar to a probe signature and thus find images that had content similar to the probe image.

The reduction of images to signatures using the PCNN and ICM have been proposed for some time. This chapter will explore the current state of this research.

6.1 Image Signature Theory

The idea of image signatures stems from biological research performed by McClurken et al. [59] They measure the neural response of a macaque to checkerboard style patterns. The brain produced neural patterns that were small and indicative of the input stimulus. They also used colour as the input stimulus and measured the colour response. Finally, a colour pattern stimulus led to a signature that was the multiplication of the pattern signature and the colour signature.

Converting images to small signatures would be of great benefit to digital image searches for two reasons. The first reason is that images do consume memory resources. JPEG compression provides a reduction of about a factor of 10. While this is impressive it may be insufficient for large databases. For example, a database of 10,000 colour images that are 512×512 will still consume several gigabytes. This is manageable for today's hard drives, but it will still take time to read, decompress, and process this amount of information. Thus, image signatures would provide an efficient representation of the image data. The second reason is that image signatures would be extremely fast to process.

6.1.1 The PCNN and Image Signatures

The creation of signatures with the PCNN was first proposed by Johnson [16] In this work two objects with equal perimeter lengths and equal areas were used. Several images were created by rotating, shifting, scaling, and skewing the objects. Johnson showed that after several PCNN iterations the number of neurons firing per iteration became a repetitive pattern. Furthermore, the pattern for each shape was little changed by the input alterations. Thus, it was possible to determine which shape was in the input space by examining the repetitive integrated pulse activity.

This experiment worked well for single objects without a background, but problems awaited this approach. First, it required hundreds of PCNN iterations which were time consuming. Second, the signature changed dramatically when a background was used. It was no longer possible to determine the input object by examining the signature.

The reason for this became clear when the problem of *interference* was recognized (see Sect. 2.2.3). Interference occurs when the neurons from one object dramatically changed the activity of neurons dedicated to another object. Thus, the presence of a background, especially a bright one, would significantly alter when the on-target neurons would pulse, and, in turn, changed the signature. In Sect. 2.2.3 this phenomenon was demonstrated. The pulsing activity of the neurons on the flower were significantly changed by the presence of a background.

The solution to the interference problem was to alter the inter-neuron connectivity scheme. The ICM therefore employs a more complicated scheme in which the connections between the neurons are altered with each iteration. Now, the signatures of on-target neurons are not altered and it is much easier to determine the presence of a target from the signature.

Johnson's signature was just the integration of the neurons that pulse during each iteration,

$$G[n] = \sum_{i,j} Y_{i,j}[n]. \tag{6.1}$$

There are still several concerns with this method. The first is that if the target only filled 20% of the input space then only 20% of the signature would be derived from the signature. Thus, the target signature could be lost amongst the larger background signature. The second concern is that it is still possible for objects of different shape to produce the same signature.

The second concern was addressed by adding a second part to the signature. The signature in (6.1) represented the area of the pulsing neurons. Area does not indicate shape and since shape is important for target recognition a second part of the signature was added that was highly dependent on the shape [58] This additional component was,

$$G[n+N] = \sum_{i,j} Z\{Y[n]\}_{i,j},\qquad(6.2)$$

where N is the total number of iterations and $Z\{\}$ is an edge enhancing function. Basically, this second component would count again the neurons that were on the edge of a collective pulsing segment.

Thus, the signature for a grey scale image was twice as long as the number of ICM iterations. Usually, N ranged from 15 to 25 so at most the length of the signature was 50 integers. This was a drastic reduction in the amount of information required to represent an image. The following sections will show results from using this type of signature.

6.1.2 Colour Versus Shape

Another immediate concern is that of colour. Most photo-images contain 3 colour bands (RGB), and it is possible to build a 3-channel ICM. However, it became apparent through trials that it was not necessarily better to process the colours in this manner. The question to be asked is what is the most important part of an image? Of course, this is based on specific applications, but in the case of building a image database from generic photos the most important information is the shapes contained within the image. In this case, shape is far more important than colour. Thus, one option is to convert the colour images to grey scale images before using the ICM. The logic is that the signature would be indicative of the shapes in the image and not the colour of the shapes. However, the debate as to use colour information or not is still ongoing.

6.2 The Signatures of Objects

The ideal signature would be unique to object shape but invariant to its location in the image and to alterations such as in-plane rotation. This is a difficult task to accomplish since such a drastic reduction in data volume makes it easier to have two dissimilar objects reduce to similar signatures.

There are two distinct objects shown in Figs. 6.1 and 6.2. The signatures for these two objects were computed independently using (6.1) and (6.2) but they are plotted together in Fig. 6.3. The plot with the square boxes belongs to Fig. 6.2. Since (6.1) and (6.2) are independent of location or rotation then neither shifting the image nor rotating the image will alter the signature by an appreciable amount.

96   6 Image Signatures

Fig. 6.1. An input image   **Fig. 6.2.** An Image

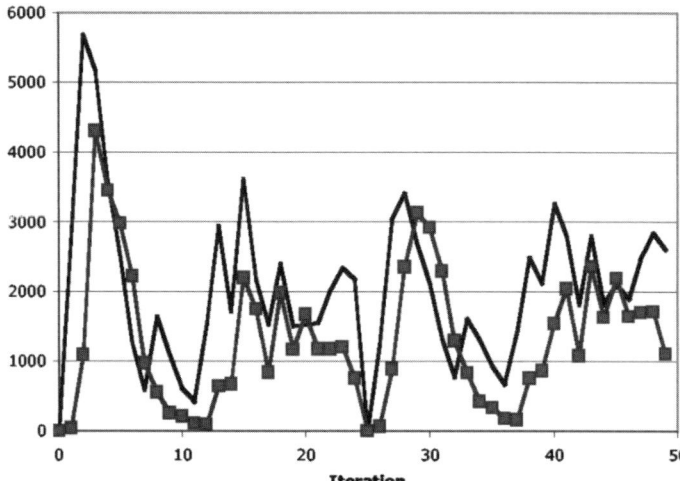

Fig. 6.3. The signatures of the two objects. The signature with the boxes corresponds to the image in Fig. 6.2

Combining the two objects into the same image will create a signature that is the summation of the signatures of the two objects. The plots in Fig. 6.4 display the summation of the two signatures in Fig. 6.2 and the signature from an image that contains the two objects.

As can be seen the signature of the two objects in the same image is the same as the summation of the two signatures. This may seem trivial but it is an important quality. For if this condition did not hold then attempts at

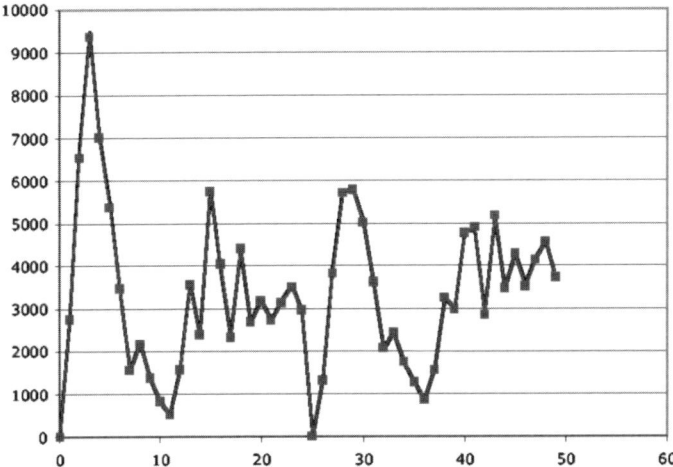

Fig. 6.4. The summation of the plots in Fig. 6.3 and the signature of a single image containing both objects of Fig. 6.1 and 6.2

target recognition would be futile. This was the case in the original PCNN signatures.

6.3 The Signatures of Real Images

The presence of a background was debilitating for the original signature method. So far the new signature method has eliminated the presence of interference, but it still remains to be seen if it will be possible to recognize a target in a real image. The image in Fig. 6.5 is the 'background' and the image in Fig. 6.6 displays the vendor pasted on top of the background.

To determine the capability of identifying a target the signature of these two images is considered. The philosophy is that the signatures of different targets is additive. Thus, G[photo] = G[background] + G[vendor] − G[occluded background]. It is expected that the signature of the vendor added to the signature of the background should almost be the same as the signature of the image in Fig. 6.6. The difference will be the portion of the background that is occluded by the target.

The chart in Fig. 6.7 displays the signature of the vendor and the difference of the signatures of the two images in Figs. 6.5 and 6.6. The target can be recognized if these two plots are similar. In cases where the background is predictable it is possible to estimate G[occluded background].

98   6 Image Signatures

Fig. 6.5. Background figure

Fig. 6.6. The vendor on the background

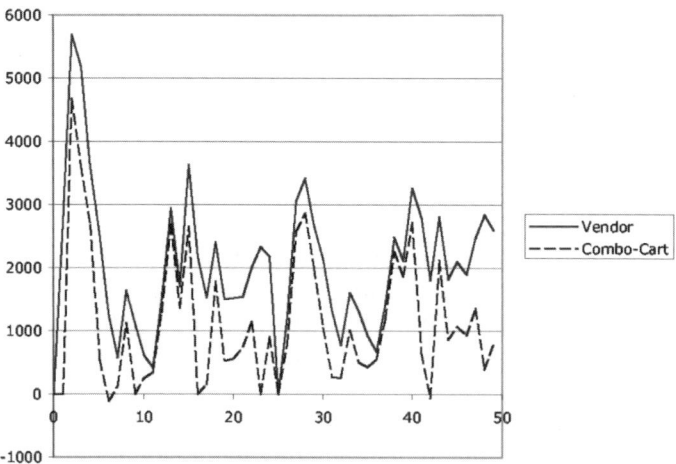

Fig. 6.7. The signature of the vendor (Fig. 6.1) and the difference of the signatures of the images in Figs. 6.5 and 6.6. The similarity of these two plots indicate that it is possible to identify the target

6.4 Image Signature Database

Another application of image signatures is to quickly search through a database of images. The signatures are far easier to compare to each other than the original images since the volume of data is dramatically reduced. Furthermore, the signatures can be compared by simple algorithms (such as subtraction) and the comparison is still independent of shift, in-plane rotation, and limited scaling. A comparison algorithm with the same qualities operating on the original images would be more complicated.

A small database of 1000 images from random sites was created. This provided a database with several different types of images of differing qualities. However, since some web pages had several images dedicated to a single topic the database did contain sets of similar images. There were even a few exact duplicates and several duplicates differing only in scale. The only qualifications were that the images had to be of sufficient size (more than 100 pixels in both dimensions) and sufficient variance in intensity (to prevent banners from being used).

The database itself, consisted of the signatures, the original URL, and the data retrieved as it was not necessary to keep the original images. Each image thus required less than two hundred bytes except for cases of very lengthy URLs.

Comparison of the signatures was accomplished through a normalized subtraction. Thus, the scalar representing the similarity of two signatures G_q and G_p was computed by,

$$a = 1.0 - \sum_n |(\|G_p[n]\| - \|G_q[n]\|)| . \tag{6.3}$$

The signatures were normalized to eliminate the effects of scale.

Comparisons of all possible pairings were computed. Since there were 1000 images in the database there were 499,000 different possible pairings (excluding self-pairings). The top scores were found and the images were manually compared.

In the data base there eight duplicate images (each pairing scored a perfect 1.0 by (6.3)). Most of the pairings scored below 0.9 and belonged to images that were dissimilar. Table 6.1 displays the results of the matches that scored above 0.9. There 13 pairings in which but images were the same except for a scale factor. There was one pairing of the same image but the scale in the horizontal and vertical were different (1.29 and 1.43). There were several pairings of similar objects and of objects that were somewhat similar. Of these 499,000 different pairings 4 scored high and contained images that did not appear to have similarity.

It was possible to mostly separate the perfect matches, the scaled pairings, and the similar images from the rest. This, of course, was not an exhaustive study since it was time consuming.

Table 6.1. Recognition rates for image signatures

Class	Scores
Different Scales	11 examples above 0.9447, 1 each at 0.9408 and 0.9126
Different Aspect	0.9016
Similar Objects	0.9380, 0.9175, 0.9163, 0.9117, 0.9099, 0.9085, 0.9081
Somewhat Similar Objects	0.9223, 0.9123, 0.9117, 0.9098, 0.9083, 0.9077, 0.9065, 0.9062
High Scoring Mismatched	0.9433, 0.9204, 0.9120, 0.9088

6.5 Computing the Optimal Viewing Angle

Nils Zetterlund

Another application is to select an optimal viewing angle for a 3D target. For example, if we wish to place a camera along side of a road to take images of autos then we need to ask: what is the best placement for the camera? Should it look at the cars from a certain height and from a certain angle to the travelling direction?

In order to answer this question we may wish to consider all possible angles of a few vehicles. We will then have to decide on a metric to define the 'best viewing angle.' The first problem encountered is that there can be a massive volume of data. Consider a case where the camera can take images that are 480×640 in three colours. There are 921,600 pixels per image. If we rotate this object about a single axis and take a picture every 5 degrees, then there will be 72 pictures each of 900 K pixels. If this object is rotated about three axes and take a picture for every 5 degrees of rotation then there will be about 3.4e11 pixels of information. Obviously, we can not simply compare images from the different viewing angles.

Image signatures have the ability to dramatically reduce the volume of information that needs to be processed and they can be used to determine the optimal viewing angle.

For this test two artificial objects will be considered. These two vehicles are shown in Fig. 6.8. Each object was rotated about a vertical axis and images for each 5 degrees of rotation were computed. For each image the signature was computed.

In order for the signatures to be useful they must smoothly change from one angle of rotation to the next. That is each element in the signature must not change in a chaotic fashion as the angle is increased. Thus, we should be able then to predict values of the elements of the signature given a sparse sampling.

A sampling set consisted of the signatures for every 30 degrees. This set then was $G_{b,\theta,n}$ where b is the class (bus or beetle), θ is the angle of viewing,

6.5 Computing the Optimal Viewing Angle

Fig. 6.8. Digital models of two vehicles

and n is the element index. Given the G's for $\theta = 0, 30, 60, \ldots$ is it possible to predict the values of G for $\theta \neq 0, 30, 60, \ldots$? Using a Gaussian interpolation function the prediction of the intermediate G values becomes feasible. The charts in Figs. 6.9, 6.10, and 6.11 display the measured and estimated values of G for two fixed values of n. The estimation does indicate that the intermediate elements are somewhat predictable. This in turn validates the use of signatures for the representation of viewing angle information.

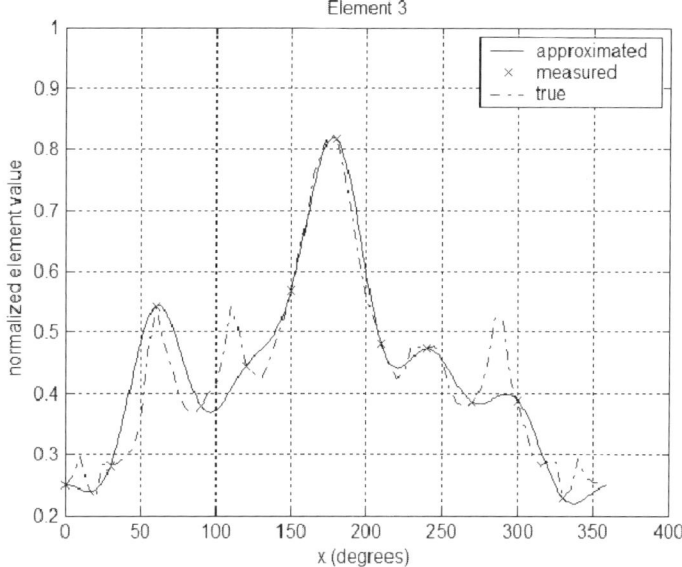

Fig. 6.9. The actual and estimated values of element 3 of the signature

102   6 Image Signatures

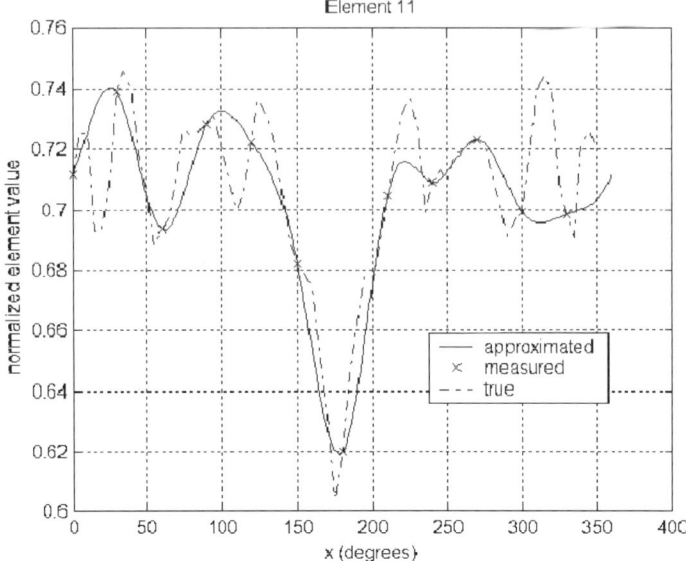

Fig. 6.10. The actual and estimated values of element 11 of the signature

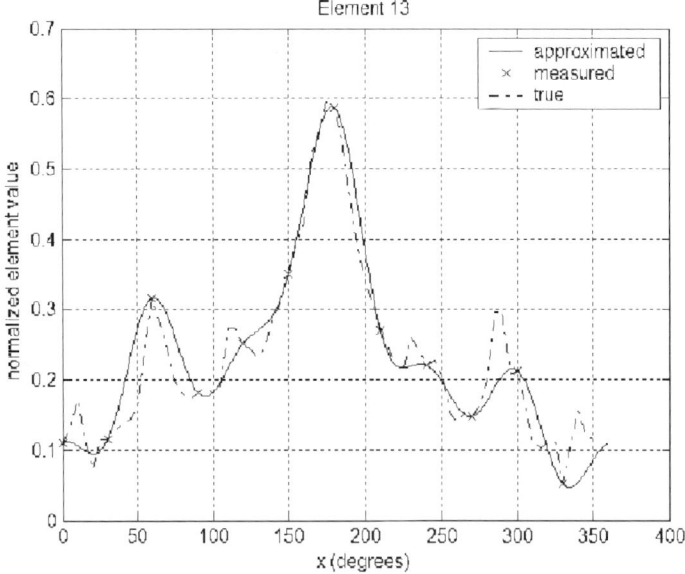

Fig. 6.11. The actual and estimated values of element 13 of the signature

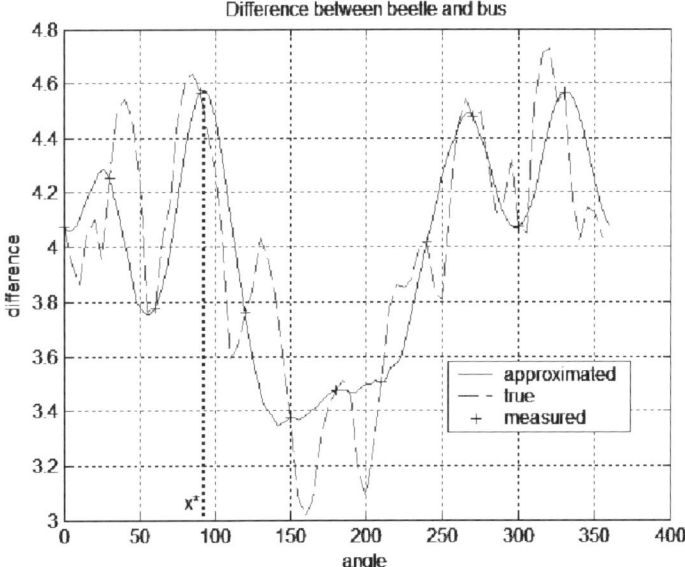

Fig. 6.12. The first order difference between the two targets

The next step in the process is to compare the signatures of two objects to find a view that is most distinguishing. In other words, we seek the viewing angle that best discriminates between the two objects. This angle will be the angle in which the signatures between the two objects differ the most. Figure 6.12 displays the first order difference between the respective signatures of the two objects. This is show for both the 5 degree and 30 degree cases. It is seen that there are in fact four angles that greatly distinguish these two objects. The first is at 90 degrees which is a side view (Fig. 6.8). The second is at 270 degrees which is the other side view. The third and fourth correspond to views at the front corners of the vehicles. Likewise, views that are directly at the front or the back of the vehicles are shown to be poor discriminators.

This is only the beginning of what signatures can do for view angle estimation. Knowing that the elements of the signatures are somewhat predictable from other measurements can directly lead to the estimation of the optimal viewing angle. In other words, if we have a few viewing angles can we predict what the best view is without having seen it before? There is still much work to be done in this area.

6.6 Motion Estimation

Image signatures have also been employed to estimate the velocity of an object. A moving input will alter the signature of a static object. The overall characteristic of the signature is maintained but small alterations are indicative of motion.

The current method of computing the signature is insensitive to opposing movement. For example, an object moving in the $-x$ direction is difficult to distinguish from the same object moving in the $+x$ direction. Thus, the signature calculation is altered to be sensitive to directionality,

$$G[n] = \sum_{i,j} Y_{i,j}[n], \tag{6.4}$$

$$G[n+N] = \sum_{i,j} (\nabla_x Y)_{i,j}, \tag{6.5}$$

and

$$G[n+2N] = \sum_{i,j} (\nabla_y Y)_{i,j}, \tag{6.6}$$

where N is the numbers of iterations (in this study $N = 25$) and ∇N_x is the spatial derivative in the x direction. Thus, the signature has three times as many elements as the number of iterations. A comparison score of two signatures (G_p, G_q) is computed by (6.3).

Figure 6.13 displays signatures from a static and moving object. The alterations to the image signature are dependent upon the speed, the direction of the motion, and the object shape.

Signatures of a moving target can be computed by (6.3). Consider an object capable of moving in 2D space with a constant velocity. We can construct a velocity space R^2 and compare the signature of all possible velocities to the signature of the static target. This comparison is shown in Fig. 6.14 where the curves depict similar values of the velocity difference Dv.

These curves are named iso-Dv since they depict constant values of the velocity difference. Of course, the *anchor velocity* does not have to be $v = 0$. We can compare all velocities in the space to any single velocity. If the iso-Dv values were solely dependent upon the difference in velocity then these curves would be circles. However, there is also a dependency upon the object shape. Thus, we expect the different iso-Dv to be similar in shape for a single target. As the Dv increases the iso-Dv curves lose their integrity, so there is an effective radius – or limit on how large Dv can be in order for (6.3) to be valid.

Now, consider the case of a target with an unknown velocity $v_?$. It is our task to estimate this velocity using image signatures. If we compare $v_?$ to a v_{x1} that is sufficiently close to $v_?$ then a value from (6.3) can be computed. However, this computation does not uniquely identify $v_?$. Rather, the computed value defines an iso-Dv curve surrounding v_{x1}. If we compare the signature from $v_?$ to two other knowns v_{x2} and v_{x3} then the three iso-Dv curves will intersect at a single location. This triangulation method is shown in Fig. 6.15. The estimate of $v_?$ is the point in R^v space where the three iso-Dv curves intersect.

6.6 Motion Estimation   105

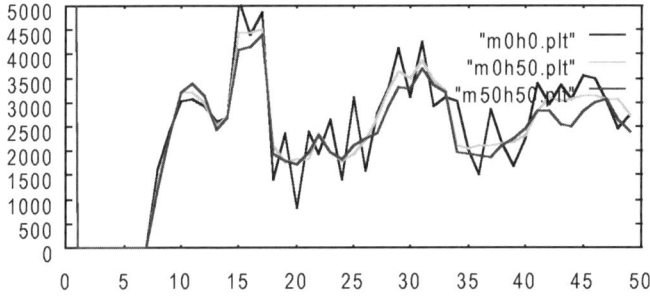

Fig. 6.13. The signature of a static object (*roughest curve*), the same object moving at velocity of $(0, 50)$ (*lightest curve*), and the same object moving at a velocity of $(50, 50)$

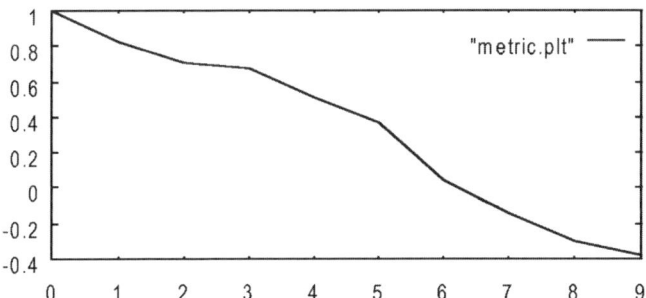

Fig. 6.14. The comparison of signatures at different velocities to the signature of the static case. The x-axis is increasing velocity and $x = 9$ is a velocity of $(0, 45)$

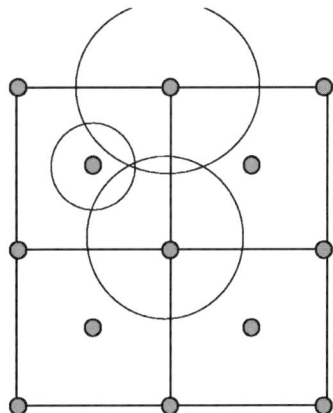

Fig. 6.15. A velocity grid with 13 anchor points. The middle point is $v = (0, 0)$. The 3 circles are the iso-Δv for an unknown. The 3 circles intersect at a single location. This location is the estimate of v

The only caveat is that v_{x1}, v_{x2} and v_{x3} must be sufficiently close to $v_?$. Since $v_?$ is unknown it is not possible to define v_{x1}, v_{x2} and v_{x3}. To circumvent this problem many anchor velocities are considered. In the trial case 13 anchor points distributed evenly in R^v were used. The signature from $v_?$ was compared to the signatures from these anchor points. The three points with the highest comparison value (6.3) were then used for the triangulation method.

For the example case R^v was defined by 100×100 points with the static v_0 defined at $(50, 50)$. The maximum velocity was one pixel of target movement per iteration of the ICM. All 10,000 points in R^v were considered as $v_?$. In all cases this method correctly estimate $v_?$ with an accuracy of $(\pm 1, \pm 1)$. For the cases in which there was an single element error the signature of $v_?$ and the correct answer were identical. Thus, the velocity was accurately predicated for all cases.

6.7 Summary

The image signatures are an efficient method for reducing the volume required to represent pertinent image information. The signatures are unique to the shapes inherent in the image and are loosely based on biological mechanics. The reduction in information volume allows for the construction of a image database that can be searched very quickly for matching images. Other uses include the determination of the optimal viewing angle and the estimation of motion.

7 Miscellaneous Applications

There is a wide variety of applications that the PCNN and ICM have proven valuable. In this chapter a few of these applications will be reviewed in a terse fashion. The PCNN has been used as a foveation engine which finds points of focus within an image. The PCNN has also been used as a logic engine solving such problems as maze-running. Finally, an application to generate a bar-code representation of an image will be presented.

7.1 Foveation

The human eye does not stare at an image. It moves to different locations within the image to gather clues as to the content of the image. This moving of the focus of attention is called foveation. A typical foveation pattern [118] is shown in Fig. 7.1. Many of the foveation points are on the corners and edges of the image. More foveation points indicate an area of greater interest.

A foveated image can be qualitatively described as an image with very rich and precise information at and around those areas of the image under intense observation, and with poorer information elsewhere. The concept of foveation, as applied here, exploits the limitations of the biological eye to discriminate and detect objects in the image that are not in direct focus. In mammalian psychophysics, most tasks are performed better with foveal vision, with performance decreasing toward the peripheral visual field. The visual acuity can change by as much as a factor of 50 between the fovea and peripheral retina.

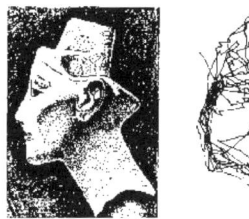

Fig. 7.1. A typical foveation pattern [12, 102, 118]

108   7 Miscellaneous Applications

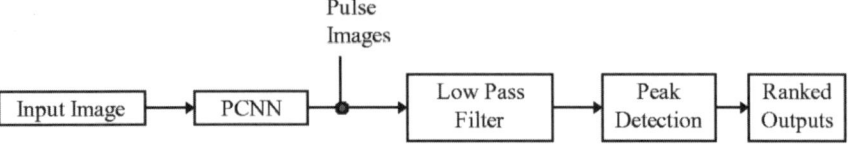

Fig. 7.2. Diagram of the logical flow of the foveation system

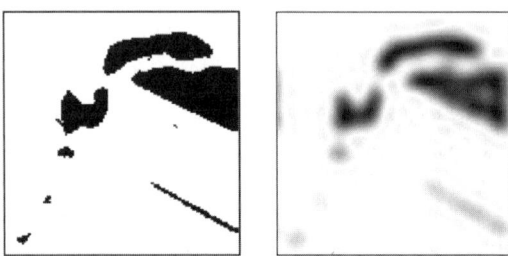

Fig. 7.3. A pulse image and the filtered version of that image. The black pixels indicate a higher intensity

7.1.1 The Foveation Algorithm

The foveation algorithm relies heavily on the segmentation ability of the PCNN. The segments produced by the PCNN are then filtered to extract the foveation points. The logical flow of the system is shown in Fig. 7.2. Basically, the PCNN produces a series of binary pulse images, which contain a few segments from the original image, and each image contains a different set of segments. These segments are sent through a low pass filter that enhances the desired areas. Examples of a pulse image and its filtered image are shown in Fig. 7.3. The filtering causes the corners and some of the edges of the larger areas to become significantly brighter than the interior. Medium sized areas will also become smooth areas with single peaks, while smaller areas will decrease to a level of insignificance. Finding the foveation areas now becomes a matter of peak detection, which is quite simple since each of the peak areas is quite smooth.

Each image was searched for peaks that were within 90% of the initial peak value. These were then reserved for further processing, which will be discussed later.

The first example involves handwritten letters. Traditional foveation points for these letters are corners and junctions. Figure 7.4 shows the original letters and the foveation points found by the PCNN. These points are ranked to indicate the order in which the foveation points were found.

The second example is that of the face from Fig. 7.1. Unfortunately, the PCNN does not work well with binary inputs and the original face image is binary. So this image was smoothed to give it texture which destroyed some of the smaller features such as the eye. The foveation points selected

7.1 Foveation   109

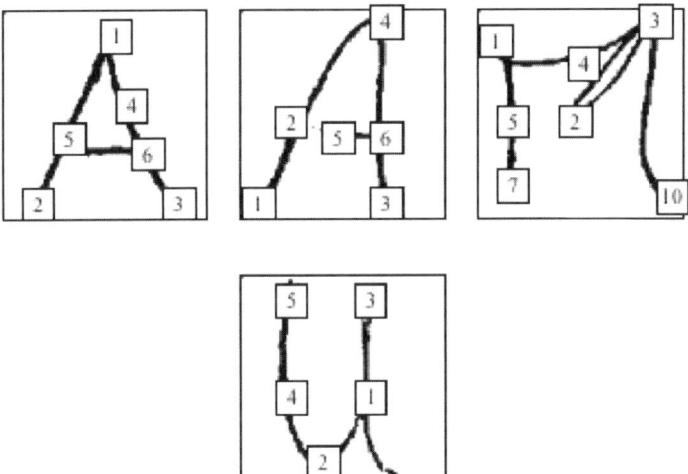

Fig. 7.4. Handwritten characters and their foveation points as determined by the PCNN-based model discussed here

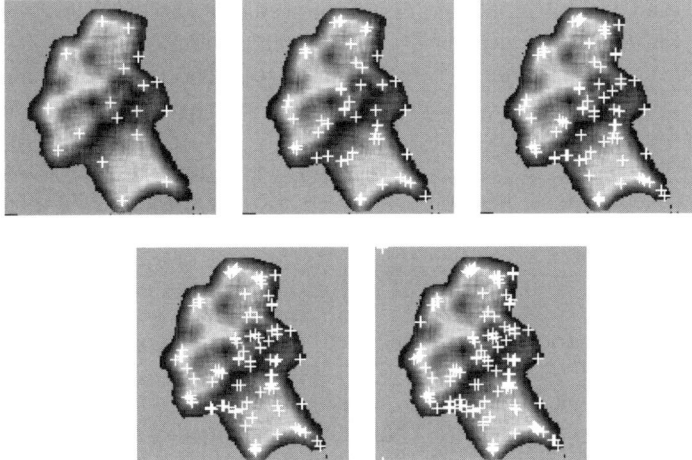

Fig. 7.5. The foveation points for the face image for 10, 20, 30, 40 and 50 iterations

by the PCNN model are shown in Fig. 7.5. The foveation patterns here are roughly similar to those in Fig. 7.1. It should also be noted that the PCNN algorithm is not simulating all of the sources for foveation. Humans are 'hardwired' to recognise humans and foveation on an image of a face is driven by more than just the shapes contained within the image. However, there are encouraging similarities between some of the human foveation points and the points computed by the PCNN model.

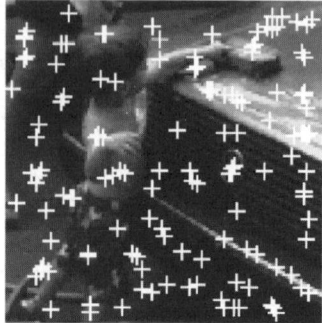

Fig. 7.6. Foveation points overlaid on a 'real-world' image

The final example is to compute the foveation points for a fairly complicated image with many objects, noise, and non-uniform background. The image and the foveation points are shown in Fig. 7.6. Many of the foveation points were along lines and edges within the image. However, less important details, such as the car grille and the features on the boy's shorts, did not produce foveation points. It should also be noted that low contrast larger features (such as the edges between the bottom of the car bumper and the ground) did produce foveation points. Thus, it can be seen that these foveation points are quite different than those that would have been produced by conventional edge filtering of the original image.

Unfortunately, it is not possible to say which foveation points are correct – or at least which ones mimic humans. However, the PCNN model does produce foveation points at the desired places, i.e. in regions of corners and edges.

7.1.2 Target Recognition by a PCNN Based Foveation Model

A test consisting of handwritten characters demonstrates the ability of a PCNN-based foveation system. The PCNN generates foveating points which are now centers of attention – or possible centers of features. The features of these centers can be identified, and using a fuzzy scoring algorithm [107] it is possible to identify handwritten characters from an extremely small training set [108].

Typical handwritten characters are shown in Fig. 7.7. In this database there were 7 samples from a single person and 1 sample of 3 letters (A, M and U) each from 41 individuals. Typical foveation patterns of these letters are shown in Fig. 7.4. The logical flow of the recognition system is shown in Fig. 7.8.

Once the foveation points are produced, new images are created by a barrel transformation centred on each foveation point. Examples of the letter 'A' and barrel transformations centred on the foveation points are shown in

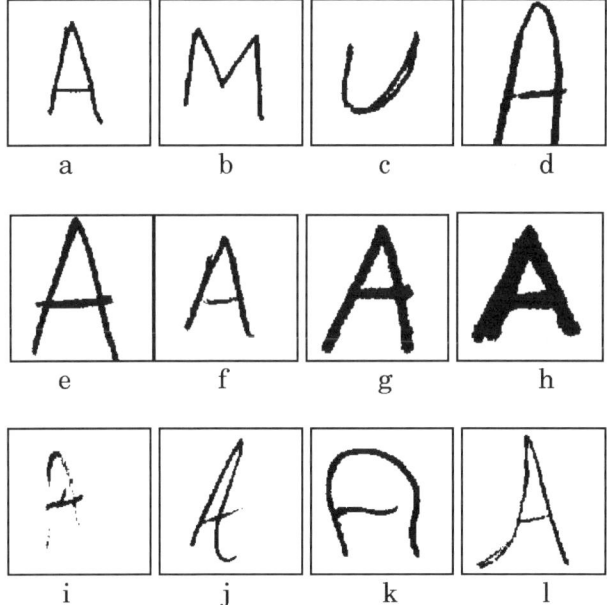

Fig. 7.7. Typical handwritten letters

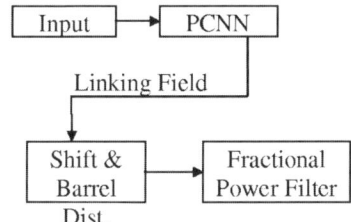

Fig. 7.8. The logical flow of the recognition system

Fig. 7.9. This distortion places more emphasis on the information closer to the foveation point. Recognition of these images constitutes the recognition of a set of features within the image and combining the recognition of these features with the fuzzy scoring method.

Recognition of the feature images is performed through a Fractional Power Filter (FPF) [26]. This filter is a composite filter that has the ability to manipulate the trade-off between generalization and discrimination that is inherent in first order filters. In order to demonstrate the recognition of a feature by this method an FPF was trained on 13 images of which 5 were target features and 8 were non-target features. For this example one target feature is the top of the 'A' (see Fig. 7.9b) and the non-targets are all other features.

112   7 Miscellaneous Applications

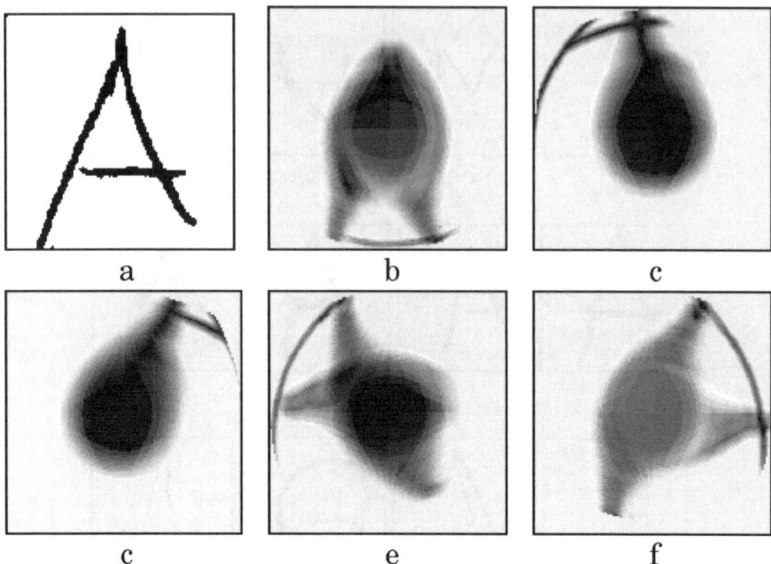

Fig. 7.9. An original 'A' and the 5 barrel distorted images based upon natural foveation points

The results of the test are presented as three categories. The first measures how well the filter recognised the targets, the second is how well the system rejected non-targets, and the third considers the case of a non-target that is similar to the target (such as the 'M' having two features similar to the top of the 'A'). The maximum correlation signature about the area of the foveation point was recorded. The FPF was trained to return a correlation peak of 1 for targets and a peak of 0 for non-targets. The results for (non-training) targets and dissimilar non-targets are shown in Table 7.1. Similar non-targets produced significant correlation signatures as expected. Certainly, a single feature cannot uniquely characterize an object. The similar features of the 'M' also produced significant correlation signals. This indicates the obvious: that single features are insufficient to recognize the object.

The results demonstrate an exceedingly good separation between targets and non-targets. There were a few targets that were not recognized well. Such target comes from Fig. 7.7i, j and k. Figure 7.7i is understandable since the object is poorly represented. Next, Fig. 7.7j performed poorly since the

Table 7.1. Recognition rates

Category	Average	Low	High	Std. Dev.
Target	0.995	0.338	1.700	0.242
Non-Target	0.137	0.016	0.360	0.129

top of the 'A' is extremely narrow. Furthermore, the "A" in Fig. 7.7k has an extremely rounded feature on top. These last two features were not represented in the training features. All of the different types of 'A's discussed produced a correlation signature above 0.8 which is clearly distinctive from the non-targets.

A few false negatives are not destructive to the recognition of the object. Following the example of [107], a collection of recognized features can be grouped to recognize the object. Noting the locations of the correlation peaks in relation to each other performs this. A fuzzy score is attached to these relationships. A large fuzzy score indicates that features have been identified and are located in positions that are indicative of the target.

It has been shown that the PCNN can extract foveation points, that attentive (barrel-distorted) images can be created and centred about foveation points. Furthermore, it has been shown that these images which now represent a feature of the image, can be recognized easily, and it has been shown elsewhere that a combination of these recognized features and their locations can be combined in a fuzzy scoring method to reach a decision about the content of the input.

7.2 Histogram Driven Alterations

One interesting attribute of the PCNN is its effect on the histogram of an averaged output image. In this scenario the PCNN runs for several hundred iterations. The outputs of each iterations are summed into a single image and normalised by the number of iterations. The resulting image looks very much like the original. However, the histogram of the averaged image is unusual. We find that most of the energy in this image lies in only a few intensity levels. Initially, the number of levels in the histogram is the number of image cycles. A cycle is roughly defined by each neuron pulsing once. For any non-zero input, all neurons will eventually pulse thus completing a cycle. Unfortunately for the study of cycles, they tend to severely overlap as the iterations progress.

The colour photo (Fig. 4.1) of the boy eating ice cream is again used as an input image. The histogram of the three colour channels of the original image is shown in Fig. 7.10. In Fig. 7.11 we have included the averaged outputs after 100 iterations and Fig. 7.12 displays the histogram of these three channels. It is interesting to note that the distribution of the averaged outputs takes on the shape of the bell curve, which is far different than the original image. Figure 7.13 displays the averaged output after 1000 iterations. The histogram of this image is shown in Fig. 7.14. We note that the distribution now has separated into just a few major bands. A majority of the information of this image is now stored in these bands.

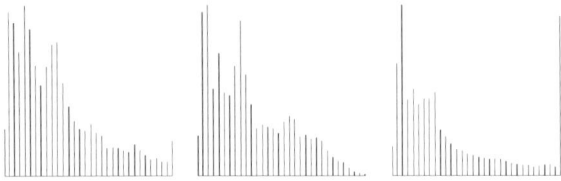

Fig. 7.10. Histogram of the RBG for the input image (Fig. 4.1)

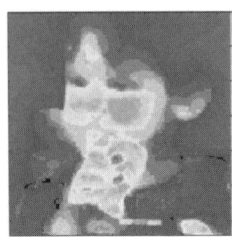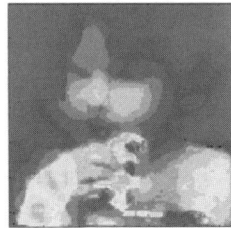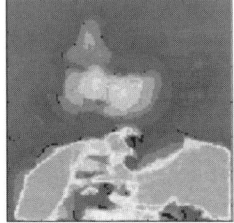

Fig. 7.11. Averaged RGB Output after 100 iterations

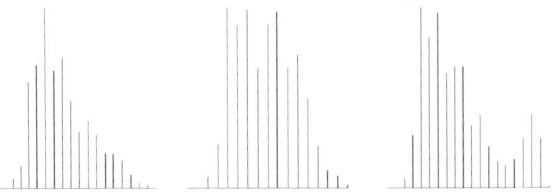

Fig. 7.12. Histogram of the RGB channels after 100 iterations

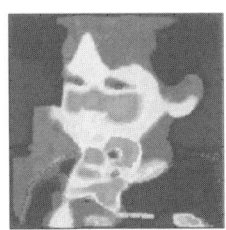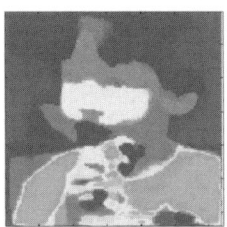

Fig. 7.13. Averaged RGB output after 1000 iterations

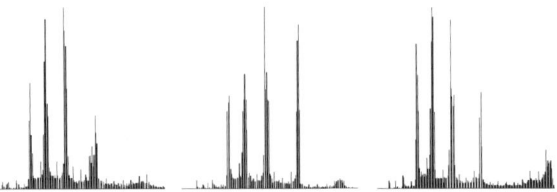

Fig. 7.14. Histogram of the RGB channels after 1000 iterations

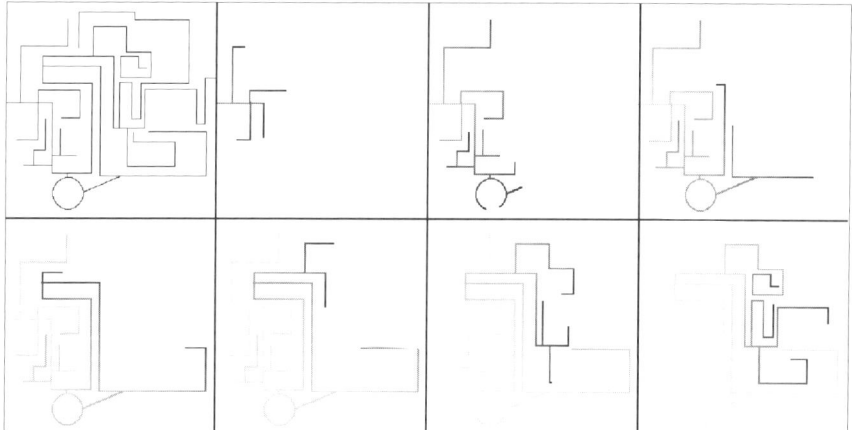

Fig. 7.15. The maze problem solved by a PCNN

7.3 Maze Solutions

Maze problems are solved by the PCNNs in a very efficient manner. There are no changes to the PCNN algorithm itself in the maze solution. A maze is constructed where the paths are X elements and all pixels off path are 0. The starting point of the maze is set to a value greater than X so that it pulses first. The PCNN is iterated and the autowave will travel the paths. X is a value that by itself will not pulse the neurons, but it will pulse the neurons when a neighbour pulses. The values of the elements of the threshold array should also be initially larger than zero. In order to find the shortest path, we simply collect each pulse output (Y) and accumulated them weighted by an increasing factor gamma. *The weighted time average allows for easy traceback by following the path of decreasing values.* The shortest path is shown in the accumulation as a monotonically increasing path. Starting from the end point one can follow the monotonically decreasing path back to the beginning along only a single path. All other paths fail the monotonic behaviour. The number of computations is solely dependent upon the shortest path length and not dependent upon the complexity of the problem. An example is shown in Fig. 7.15. The upper left image is the maze with the starting point to the left (middle). and the goal to the right. The next image shows the sum of 50 PCNN temporal (binary) outputs each tagged by the aforementioned gamma-weights as discussed above, yielding a grey scale along the path. The next image includes the 100 first outputs, etc. Note that the gamma tagging yields a darker grey scale closer to the end. The last picture (lower right corner) ends close to the exit with the darkest shade of grey. Tracing the continuously decreasing grey scale from the exit yields the path through the maze.

The solution is not limited to a maze with thin paths as shown in Fig. 7.16. A thick maze contains paths that have a significant width. Without modi-

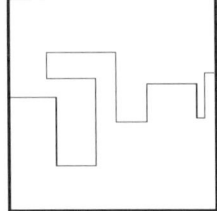

Fig. 7.16. The solution – shortest path through the maze

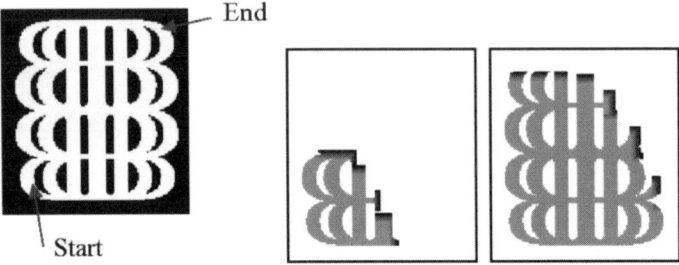

Fig. 7.17. a) shows a thick maze and **b)** and **c)** show autowaves traveling through a thick maze

fication to the algorithm the PCNN can also find the shortest path from a START to and END point. In Fig. 7.17 a thick maze is displayed with designated START and END points. The other two images display the progression of the autowaves that the PCNN generates. These follow thick paths just as easily as thin paths.

7.4 Barcode Applications

Soonil D.D.V. Rughooputh and Harry C.S. Rughooputh

The technique described in this section relates to the generation of binary barcodes from images (including spectra) and data for use in several applications. We make use here of PCNN to produce (1D/2D) binary barcodes of (2D/3D) images (including spectra) and data sequences – described in Sect. 7.4.1. A few selected applications is given in Sect. 7.4.2. The PCNN has been demonstrated to produce time signatures that are unique to the original images/data sequence. The process of barcode signature generation is based on a subsequent conversion of these time signatures into binary barcodes (1D/2D). These barcodes are unique to the input images/data and as such have many potential applications in many fields. A number of parameters are adjustable to suit individual applications that render this barcoding technique secure, versatile and robust. Applications are numerous and include image analysis and processing, recognition, diagnostics, condition monitoring, sorting and grading, classification, counting, security, cryptography, counterfeit, authenticity/signature verification, copyright, information storage and retrieval, data mining, surveillance, forecasting, feature extraction, control system applications, navigation, learning, inspection, sensor/robotic applications, change detection, defect detection, fault detection, data fusion, error detection, coding, animation and virtual reality, sequence analysis. The field of applications of concern include Space applications, Electronics, Computer Vision, Military, Surveillance, Forensic, Aids to disabled, Biomedical, Instruments and Devices, Pattern Recognition, Spectral recognition, Learning, Classification, image processing and analysis, Sensors, Communications, Internet and Multimedia, Meteorology, Digital Transmission, Coding.

7.4.1 Barcode Generation from Data Sequence and Images

We now describe the technique for the generation of barcodes from (a) Still Images and (b) Data Sequences.

Binary Code for a Still Image

The technique for the generation of binary barcode for still images requires the use of two PCNNs – PCNN#1 and PCNN#2 is shown in Fig. 7.18. PCNN#1 is used for time signature generation from the still image. This time signature is then grey-coded (8-bit) as an image. The PCNN#2 is then used for binary barcode generation from the grey-coded time signature. Alternatively, PCNN#2 can be used to obtain a time signature output which is then fed back (dotted lines in Fig. 7.18a) to obtain the corresponding grey-coded image. This feedback can be iterated any number of times and the parameters of PCNN#2 can be varied for individual iteration.

Algorithm

A.1 Time Signature Generation from Still Image

Step 1:
Set Parameters for the PCNN#1 (Decay parameters, thresholds, potentials, Number of Iterations).

Step 2:
Present Image (Fig. 7.18b) (direct or pre-processed) to PCNN#1 to generate time signature (also referred as ICON) – Fig. 7.18c. For a particular set of parameters for the PCNN#1, there is a 1:1 correspondence between the input to the PCNN#1and the output icon.

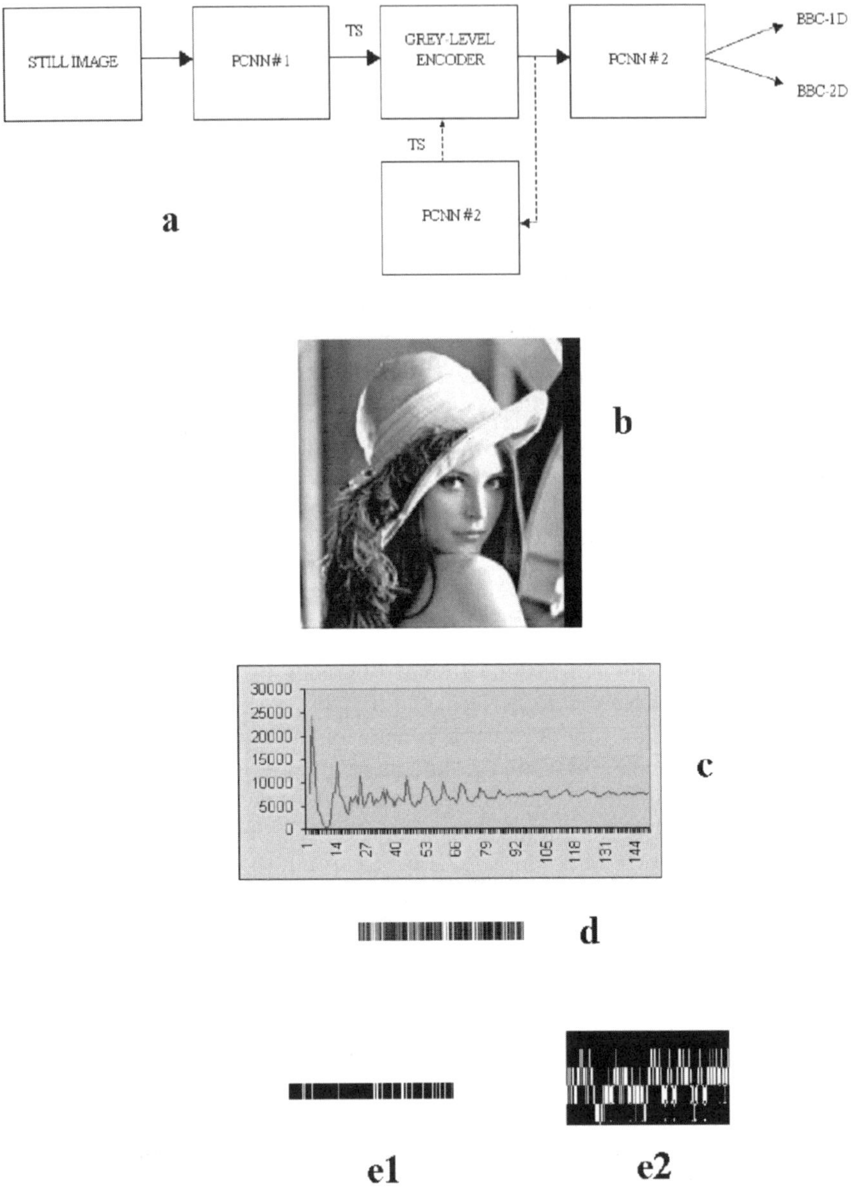

Fig. 7.18. Barcode generation from still images. **a)** Block diagram of barcoding technique for a still image, **b)** still image, **c)** time signature of (**b**), **d)** 1D binary barcode, **e)** 2D binary barcode

A.2 Grey Level Barcode (GBC) Generation from ICON

Step 3:
Set Number of Colour Levels and Select order of Colour Levels.
The set order determines the end-use sensitivity.
Convert ICON to grey-level barcoded image (Fig. 7.18d).

A.3 Binary Barcode Generation (BBC) from GBC

Step 4:
Set Parameters for the PCNN#2 (Decay parameters, thresholds, potentials, Number of Iterations N)
(if $N = 1$ go to step 3a, else step 3b).

Step 5a:
Present Image (Fig. 7.18d) to PCNN#2 to generate binary barcode for the first iteration to obtain a 1D BBC (Fig. 7.18e1). For a particular set of parameters for the PCNN#2, there is a 1:1 correspondence between the input to the PCNN#2 and this output binary barcode.

Step 5b:
Present Image (Fig. 7.18d) to PCNN#2 to generate binary barcode for 2D BBC (Fig. 7.18e2). For a particular set of parameters for the PCNN#2, there is a 1:1 correspondence between the input to the PCNN#2 and the output binary barcode set.

Note: Step 4': Alternatively in Step 4, the parameters of PCNN#2 can be set to produce a time signature of Fig. 7.18d. This time signature is fed back for greycoding (Back to Step 3). This process (feedback option) can be repeated any number of times. However, the final steps will be 4–5 for BBC generation.

Binary Barcode for Data Sequences

The technique for the generation of binary barcode for a given data sequence given either as data pairs (x, y) where $(x, y) \in R$ or complex numbers $(x + jy) \in Z$, described hereunder, requires the use of one PCNN – refer to Fig. 7.19. This PCNN is used to generate the binary barcode directly from the grey-coded data sequence. Alternatively, this PCNN can be used to obtain a time signature output that is then fed back (dotted lines in Fig. 7.19a) to obtain the corresponding grey-coded image. This feedback can be iterated any number of times and the parameters of the PCNN can be varied for individual iteration.

Algorithm

B.1 Grey Level Barcode (GBC) Generation from Data Sequence

Step 1:
Set Number of Colour Levels and Select order of Colour Levels.
The set order determines the end-use sensitivity.
Convert Data Pair Sequence (Fig. 7.19b) to grey-level barcoded image (Fig. 7.19c).

120   7 Miscellaneous Applications

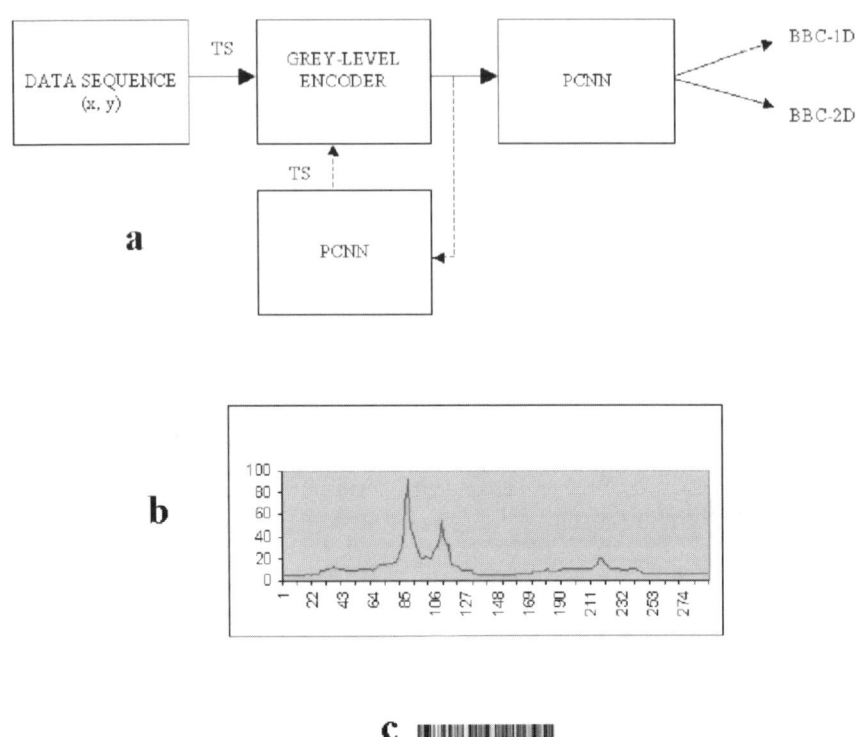

Fig. 7.19. Barcode generation from still images. **a)** Block diagram of barcoding technique for a still image, **b)** data sequence as an x,y plot, **c)** grey level barcode of (**b**), **d**) 1D binary barcode, **e**) 2D binary barcode

B.2 Binary Barcode Generation (BBC) from GBC

Step 2:
Set Parameters for the PCNN (Decay parameters, thresholds, potentials, Number of Iterations N)
(if $N = 1$ go to step 3a, else step 3b).

Step 3a:
Present GBC (Fig. 7.19c) to PCNN to generate binary barcode for the first iteration to obtain a 1D BBC (Fig. 7.19d1).

Step 3b:
Present GBC (Fig. 7.19c) to PCNN to generate binary barcode for 2D BBC (Fig. 7.19d2). For a particular set of parameters for the PCNN, there is a 1:1 correspondence between the input to the PCNN#2 and the output binary barcode set.

Note: Step 2': Alternatively in Step 2, the parameters of PCNN can be set to produce a time signature of Fig. 7.19c. This time signature is fed back for greycoding (Back to Step 1). This process (feedback option) can be repeated any number of times. However, the final steps will be 2–3 for BBC generation.

7.4.2 PCNN Counter

The barcoding method described above means that it well suited for counting objects. We illustrate this in Fig. 7.20 by applying the technique to a set of identical or mixtures of objects (closed rectangles and/or ellipses in the figure). The results clearly demonstrate the usefulness of this technique for the purpose – a one-stage process irrespective of the position of the objects. We note that, as long as the objects have non-overlapping positions, the resulting barcodes are the same.

7.4.3 Chemical Indexing

There exists several ways to search for chemical information from the Internet based World-Wide-Web system using a web browser [87]. Examples of WWW-based chemical search server/engines include Chemical Abstracts Services, ChemExper Chemical Directory, ChemFinder WebServer (CambridgeSoft), NIST database, ChemIDplus (Specialized Information Services), Hazardous Substances Databank Structures (HSDB) (Specialized Information Services), NCI-3D (Specialized Information Services), General-purpose databases of WWW contents (such as Yahoo and Alta Vista). WebServer capacities range from several thousands to several millions with databases varying from general to specialized chemicals such as liquid crystals, pesticides, polyclinic aromatic hydrocarbon, drugs, environmental pollutants, potential toxins etc. Most of these databases are freely accessible to researchers from academic and industrial laboratories.

One can search various electronic databases for chemicals by their chemical names (some accept wildcards and/or typographic variations in names) – CA index, IUPAC names, common names, trade names or synonyms, molecular formula, molecular weight, Chemical Abstracts Service (CAS) Registry Numbers (CAS RNs are unique identifiers for chemical compounds with standard format being xxxxxx-xx-x), catalog number, chemical characteristics, 2D chemical structures and substructures and molecular descriptors. The databases will identify the type of search you want, and provide the hits accordingly. Today's chemical databases are more versatile, becoming faster and faster, and corrects for obvious errors as well as invalid CAS RNs.

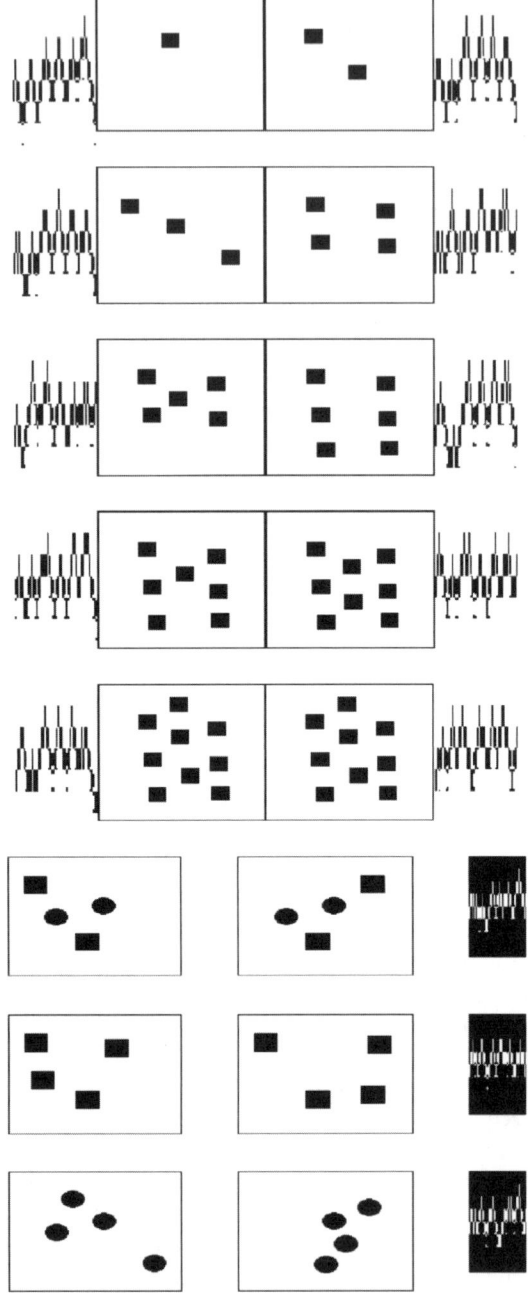

Fig. 7.20. PCNN Counter

Although sparse for the time being, some of the chemical databases also provide additional information such as 2D/3D chemical structures (as Windows metafiles or molfiles) and useful references. Helper applications or Viewers are normally needed to display chemical structural records of the compounds. Web Browsers cannot read these without a helper application and the appropriate plug-ins. In order to display a structure, a structure drawing program or WWW viewer must be used. Examples of software for chemical structures/viewers include ChemDraw, Chem3D, ChemOffice and ChemOffice Pro from CambridgeSoft, ISIS/Draw (MDL Information Systems, Inc.), Wetlab (Molecular Simulations, Inc.), ChemWeb (Softshell International, Ltd.), Accord Internet Viewer (Synopsis Scientific Systems) and Rasmol viewer. It should be noted that most of the databases are not user-friendly and often require hours of training. Besides, one normally finds that not all structures in the database currently have chemical formulas / molecular weights assigned.

Quite often one would like to identify a particular compound or a related compound (simple or complex) directly from its chemical structure alone without a priori knowledge of the CAS number, molecular formula, chemical functionality details and so on. Recognition of chemical structures can normally be a slow process requiring, in most cases, electronic submissions to the Server.

Traditionally searches are carried out as follows. Given a chemical structure such as the one shown in Fig. 7.21, we are interested to retrieve the chemical information on this compound. The traditional method, in the event a name-cannot be generated, a molecular formula is computed from the structure. Then, the elements are arranged in *Hill order*. For carbon-containing compounds, the Hill order means that carbons are listed first, hydrogens are listed next, and then all other elements are listed in alphabetical order. For compounds that do not contain carbon, the elements are arranged in alphabetical order and the number of each element is indicated. A molecular formula search for this compound is then carried out. Obviously, there may be many chemical structures that may satisfy the molecular formula provided. Thus, there may be a need for further refinement. Table 7.1 shows a typical chemical search result using the above chemical structure. In this example, a

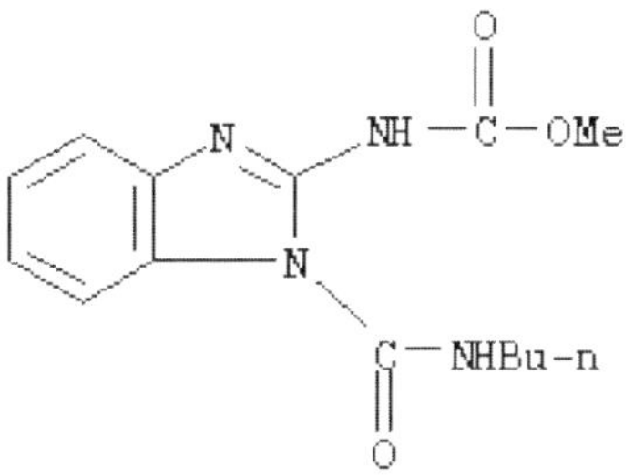

Fig. 7.21. A chemical structure

search using the molecular formula shows that there are many possible candidates. To reduce the number of answers from the molecular formula search, the latter is combined with name fragments in the Basic Index of the Registry File. The process is continued until a smaller set is obtained. Then the display option can be used to identify the RN and the name of the chemical (RN: 125705-88-6, Index Name: Carbamic acid, [5-[(3-methyl-1-oxobutyl) amino]-1H-benzimidazol-2-yl]-,methyl ester).

Current Searching Technique

The traditional method of retrieving the chemical information on a compound such as in Figure III, in the event the name is either not familiar or cannot be generated, a molecular formula is first obtained from the chemical structure. The elements therein are then arranged in the so-called Hill order. For carbon-containing compounds, the Hill order means that carbons are listed first, hydrogens are listed next, and then all other elements are listed in alphabetical order. For compounds that do not contain carbon, the elements are arranged in alphabetical order and the number of each element is indicated. A molecular formula search for this compound is then carried out. Obviously, there may be many chemical structures that may satisfy the molecular formula provided. Thus, there may be a need for further refinement. Table 7.2 shows a typical chemical search result using the chemical structure in Fig. 7.21. In this example, a search using the molecular formula shows that there are many possible candidates. To reduce the number of answers from the molecular formula search, the latter is combined with name fragments in the Basic Index of the Registry File. The process is continued until a smaller set is obtained. Then the display option can be used to identify the RN and the name of the chemical (RN: 125705-88-6, Index Name: Carbamic acid, [5-[(3-methyl-1-oxobutyl) amino]-1H-benzimidazol-2-yl]-,methyl ester).

The traditional method can thus be very time-consuming and tedious to the casual user. Searches on chemical databases are presently dominated by the text-based content of a paper that can be indexed into a key-word searchable form. Such traditional searches can prove to be very time consuming and discouraging to the less frequent scientist.

We can make use of the PCNN to produce binary barcodes of images of chemical structures [101]. A number of parameters are adjustable to suit individual applications; renders this barcoding technique secure, versatile and robust. In Fig. 7.22 we illustrate the invariance of the technique to translation (A,B) rotation (A, C, D) scale (A,E) using a thiophene molecule. The uniqueness property is sensitive to the drawing specification format as shown in (A,F) where the ring size and the bond lengths have been altered. Thus, it is important that all chemical structures are drawn according to a standard entry format. In Fig. 7.23 we show the molecular structures of a number of diverse chemical structures and their corresponding binary barcodes. The uniqueness of the chemical structures and the binary barcodes suggest

Table 7.2. A chemical structure

Molecular Formula Search

$\Rightarrow$ *FILE REGISTRY*

$\Rightarrow$ *C14H18N4O3/MF*

E1 1 C14H18N4O2TL/MF
E2 1 C14H18N4O2ZR/MF
E3 248 $\rightarrow$ C14H18N4O3/MF
E4 1 C14H18N4O3.(C2H4O)NC15H24O/MF
E5 1 C14H18N4O3.(CH2O)X/MF
E6 1 C14H18N4O3.1/2CL4PT.H/MF
E7 1 C14H18N4O3.1/2H2O4S/MF
E8 1 C14H18N4O3.2C2H6O3S/MF
E9 1 C14H18N4O3.2C7H3IN2O3/MF
E10 1 C14H18N4O3.2C7H6O2/MF
E11 3 C14H18N4O3.2CLH/MF
E12 3 C14H18N4O3.BRH/MF

Molecular Formula/Fragment Search

$\Rightarrow$ *S E3 AND BUTYL*
        248 C14H18N4O3/MF
        526169 BUTYL
L1 28 C14H18N4O3/MF AND BUTYL

$\Rightarrow$ *S L1 AND AMINO AND METHYL*
        1927517 AMINO
        6665678 METHYL
L2 10 L1 AND AMINO AND METHYL

---

that this technique can be easily exploited for direct structure recognition of chemicals. The binary barcodes generated from chemicals forms a database. Comparing its binary sequences to those in a database can then search the barcode of a chemical structure. Because of the 1:1 correspondence, the hit from the database is thus a one step procedure. The method is thus direct, less cumbersome (compared with traditional methods) and proves to be robust, elegant and very versatile; and can be considered as a serious option in lieu of CAS RNs.

126   7 Miscellaneous Applications

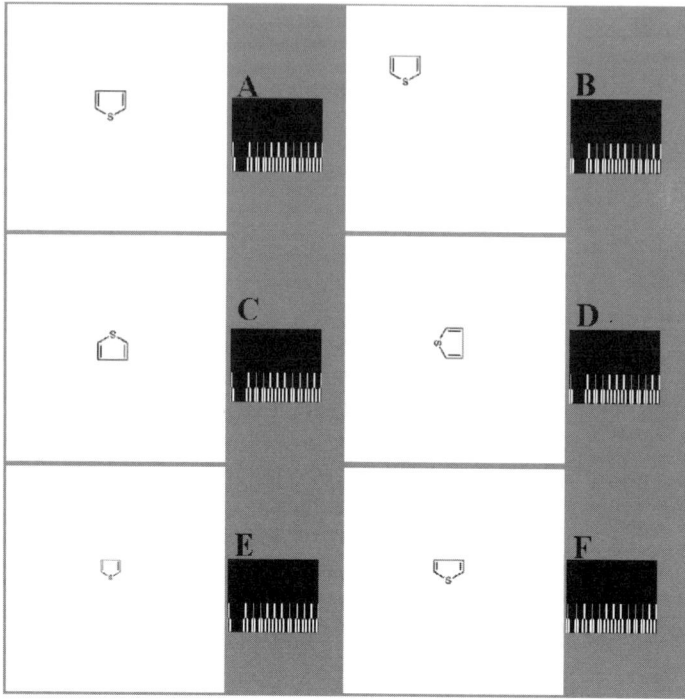

Fig. 7.22. Effects of translation, rotation, and scaling

7.4.4 Identification and Classification of Galaxies

Astronomers predict that the universe may potentially contain over 100 billion galaxies. Classification of galaxies is an important issue in the large-scale study of the Universe. Understanding the Hubble sequence can be very useful in understanding the evolution of galaxies, cosmogony of the density-morphology relation, role of mergers, understanding how normal and barred spirals have been formed, parameters that do and do not vary along it, whether the initial star-formation rate is the principal driver, etc.. Our understanding of them depends on how good the sensitivity and resolving power of existing telescopes (X-ray, Optical, Infra-red, Radio, etc.). Database construction of these galaxies has only just begun. Recognition or classification of such large number of galaxies is not a simple manual task. Efficient and robust computer automated classifiers need to be developed in order to help astronomers derive the maximum benefit from survey studies. There are different ways of classifying galaxies, for example, one can look only at the morphology. Even in this classification scheme there exists different techniques. One can combine morphology with intrinsic properties of the galaxy such as the ratio of random to rotational velocities, amount of dust and gas, metallicity, evidence of young stars, spectral lines present and their widths, etc.

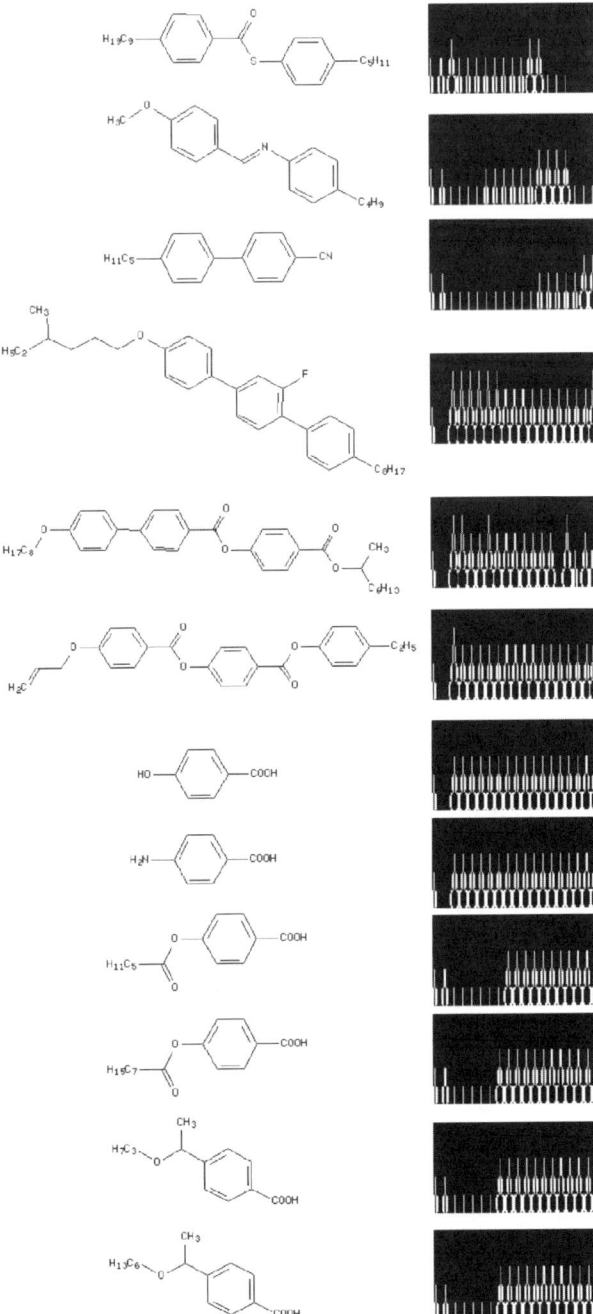

Fig. 7.23. Examples of chemical structures and their corresponding 2D barcodes

Nobody believes that galaxies look today as they did just after they were formed. Newly born galaxies probably did not fit the present-day classification continuum. The big puzzle is to find out how the galaxies evolved onto the present forms. Did the primeval galaxies "evolve" along basically the present Hubble sequence, or did they fall into the sequence, each from a more primitive state, 'landing' at their present place in the classification according to some parameter such as the amount of hydrogen left over after the initial collapse [103]?. The collapse occurred either with some regularity [70], or chaotically [104] within a collapsing envelope of regularity, or, at the other extreme, in complete chaos. Present-day galaxies show variations of particular parameters that are systematic along the modern classification sequence. The obvious way to begin to search for the 'master parameter' that governs the formation process is to enumerate the variation of trail parameters along the sequence. Reviews by [96] and by [63] indicate somewhat different results that nevertheless are central to the problem. Within each morphological type, there is a distribution of the values of each parameter, and these distributions have large overlap from class to class. Hence, the dispersion of each distribution defines the vertical spread along the ridgeline of the classification continuum (i.e. the center line of Hubble's tuning fork diagram). Hence, besides a 'master parameter' that determines the gross Hubble Type (T), there are other parameters that spread the galaxies of a given type into the continuum of L values [115]. The fact that so many physical parameters vary systematically along the Hubble sequence is strong evidence that the classification sequence does have a fundamental significance.

The easiest property of a galaxy to discuss is its visual appearance. When Hubble introduced his classification, he thought it might represent an evolutionary sequence with galaxies possibly evolving from elliptical to spiral form but, this is not believed to be true today. Hubble's classification scheme, with some modifications, is still in use today. Galaxies are classified as spiral galaxies (ordinary spirals, barred spirals), lenticulars, elliptical galaxies and irregular galaxies; including other more specialized classifications such as as cD galaxies. The spirals are classified from Sa to Sc (ordinary spirals) and from SBa to SBc (barred spirals); a to c represent spiral arms that are increasingly more loosely wound. The elliptical galaxies are classified according to their ratio of their apparent major and minor axes; the classification is based on the perspective from Earth and not on the actual shape. The lenticulars are intermediate between spirals and ellipticals. There are other classification schemes like de Vaucouleurs, Yerkes, and DDO methods, which look into higher details.

The morphological classification of optical galaxies is done more or less visually. Better classification schemes would certainly help us to know more about the formation and evolution of galaxies. A technique that involves the use of robust software to do the classification is crucial, especially if we want to classify huge number of galaxies at one shot. Since there are billions of galaxies, a robust automated method would be desirable. Several authors have reported work along this line [see Refs. 28–37 in [99]]. The techniques studied include the use of statistical model fitting, fuzzy algebra, decision tree, PCA, and wavelet-based image analysis. Some work has been reported on the use of artificial neural networks for automatic morphological classification of galaxies; using feed-forward neural network, self-organizing maps, computer vision technique – see Refs. 32–34 in [99]. These techniques, however, require extensive training, hence are computationally demanding and may not be appropriate for the classification of a large number of galaxies. A galaxy classifier/identifier using PCNN has also been reported [99,106]; initial results of which are promising. These authors have been able to classify galaxies according to an index parameter obtained from the time signature of the galaxies. The results reveal that this technique is fast and can be used for real-time classifications. The researchers have chosen a catalogue of digital images of 113 nearby galaxies [73] since these galaxies are all nearby, bright, large, well-resolved, and span the Hubble classification classes. Besides, Frei et al. photometrically calibrated all data with foreground stars removed and the catalogue is one of the first data set made publicly available on the web. Important data on these galaxies published in the "Third Reference Catalogue of Bright Galaxies" [67] are recorded in the FITS file headers. All files are available through anonymous FTP from "astro.princeton.edu"; they are also available on CD-ROM from Princeton University Press.

Binary barcodes corresponding to galaxies can be generated to constitute a databank that can be consulted for the identification of any particular galaxy (for e.g. using the N-tuple neural network). The digital image of a galaxy is first presented as input to a PCNN to produce segmented version output of binary images. Figure 7.24 shows a set of original images of representative galaxies spanning over the Hubble classification classes; corresponding NGC values are given in Table 7.3. Figure 7.25 shows the original images of representative galaxies spanning over the Hubble classification classes (from top to bottom: NGC 4406, NGC 4526, NGC 4710, NGC 4548, NGC 3184, and NGC 4449) and the corresponding segmented images for the first five iterations (column-wise). Figure 7.26 shows the set of the third iteration images for a number of galaxies listed in Table 7.3 for producing the time signatures using a second PCNN. The segmented image version was used instead of the original image to minimize adverse effects of galactic halos.

Table 7.3. NGC values for galaxies in Fig. 7.24

3184	3726	4254	4374	4477	4636	5813
3344	3810	4303	4406	4526	4710	6384
3351	3938	4321	4429	4535	4754	4449
3486	4125	4340	4442	4564	4866	4548
3631	4136	4365	4472	4621	5322	5377

130   7 Miscellaneous Applications

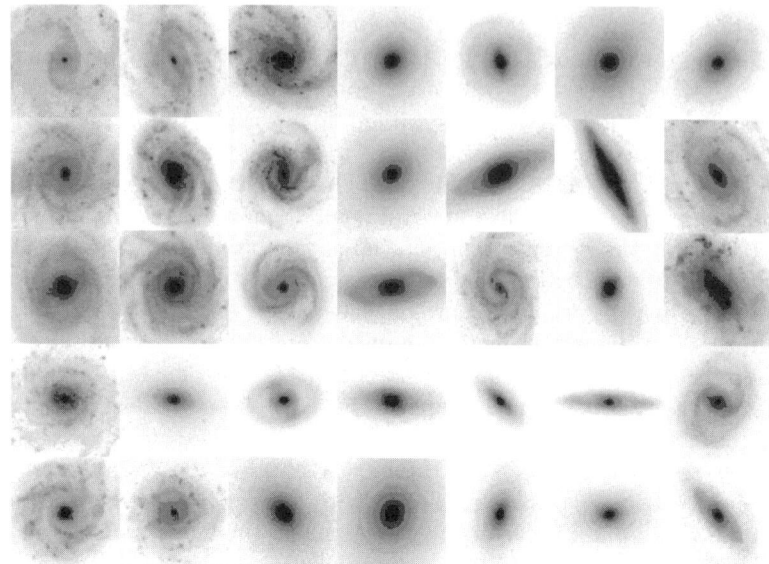

Fig. 7.24. Representative galaxies

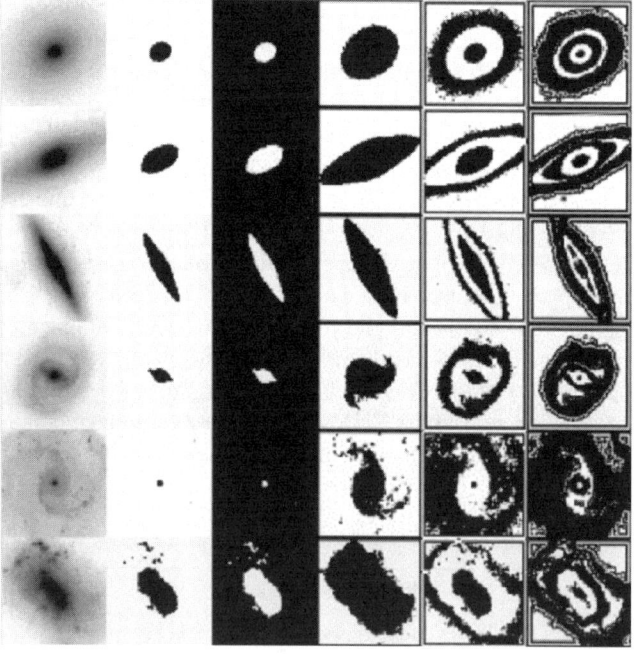

Fig. 7.25. Representative galaxies and their PCNN segmented images

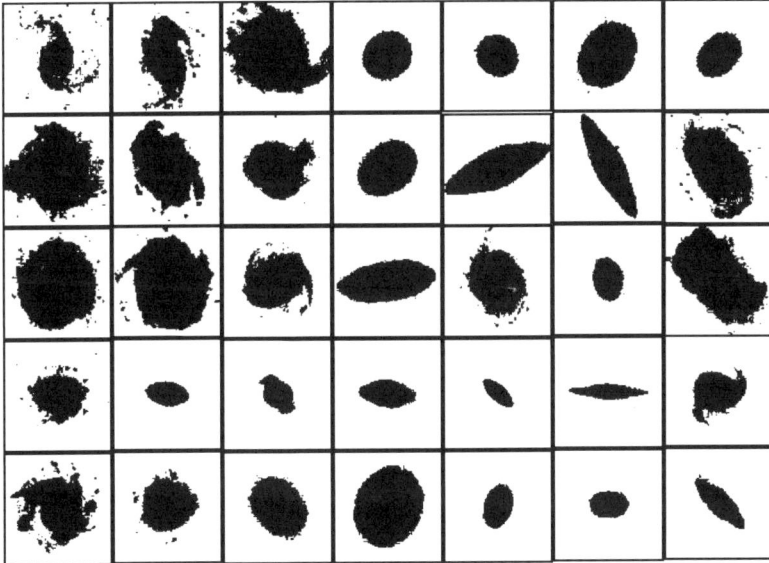

Fig. 7.26. Third iterated PCNN binary images of galaxies (see Table 7.3 for corresponding NGC values)

A method was devised to mathematically compute an morphology index parameter (mip) from the first few iterations (mip = $G(3)^2/(G(2)G(4))$) since these are related to the image textures and hence retain useful information on the morphology of the galaxies [106]. We found that galaxies (except for NGC4472) with mip values less than 10 are spirals or irregulars otherwise ellipticals or lenticular (refer to Table 7.4). This exception may be due to the presence of halos. Figure 7.27 shows the corresponding barcoded images of the galaxies listed in Table 7.3 (obtained using the corresponding. 8-bit grey level version of the time signatures). We note that the 1:1 correspondence between the barcodes and the input NGC images.

7.4.5 Navigational Systems

Considerable research has been conducted to improve safety and efficiency on navigational systems. For efficient control and increased safety, automatic recognition of navigational signs has become a major international issue. With the increasing use of semi-autonomous and autonomous systems (vehicles and robots), the design and integration of real-time operated navigational sign recognition systems have also gained in popularity. Systems that assist navigators or provide systems with computer vision-based navigational sign detection and recognition mechanisms have been devised. Manufacturers in Europe, USA, and Japan and several universities even combined their efforts in this direction. The standards used in the design of navigational signs are

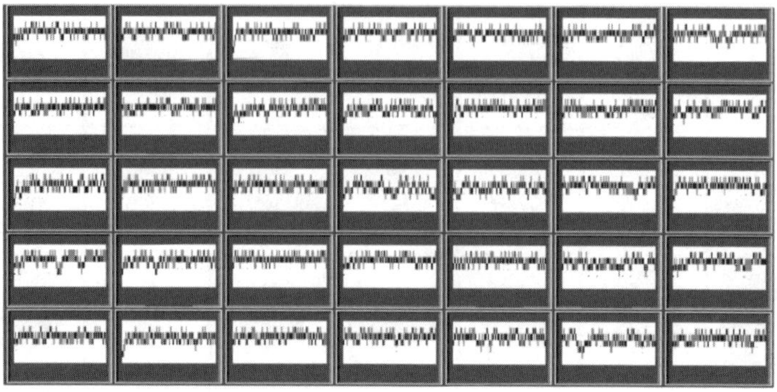

Fig. 7.27. Galaxies identified from their corresponding barcodes

Table 7.4. Galaxies of different Hubble types (T) with the mip computed

NGC	T	mip	NGC	T	mip	NGC	T	mip
3184	6	5.3	4303	4	9.2	4526	−2	11.1
3344	4	3.8	4321	4	7.2	4535	5	7.6
3351	3	4.9	4340	−1	16.3	4564	−5	18.1
3486	5	8.1	4365	−5	10.5	4621	−5	16.3
3631	5	4.6	4374	−5	13.9	4636	−5	10.9
3726	5	5.1	4406	−5	12.0	4710	−1	10.7
3810	5	6.0	4429	−1	10.7	4754	−3	15.9
3938	5	4.5	4442	−2	15.6	4866	−1	15.4
4125	−5	16.9	4449	10	5.1	5322	−5	17.3
4136	5	9.3	4472	−5	8.1	5813	−5	14.7
4254	5	4.5	4477	−3	14.9	6384	4	5.2

typically according to size, shapes, and colour compositions. These signs form a very unique and easily visible set of objects within a scene. They always appear in a visible and fairly predictable region of an image. The only significant variables are the sizes of the signs in images (due to distance) and illumination of the scene (such as bright sunlight, overcast, fog, night). Two main characteristics of navigational signs are normally used for their detection in camera-acquired images, namely colour [61,64,69,74,76,83,88–90,93,111,112] and shape [60,62,68,77,81,82,84,91,92,94,109,110]. Sign recognition is performed using sign contents such as pictograms and strings of characters. Normally, colour is employed in combination with shape for detection purposes first and then for sign recognition. Different types of image processing can be performed using colours. The three most widely used approaches are neural network-based classifiers, colour indexing, and image segmentation based on colour.

Neural network-based classifiers involve the use of neural networks specifically trained to recognise patterns of colours. The use of multi-layer neural networks as experts for sign detection and recognition has been reported and applied a neural net as classifier to recognise signs within a region of interest [83]. Swain [111, 112] have developed the 'colour indexing' technique that recognises signs by scanning portions of an image and then comparing the corresponding colour histograms with colour histograms of signs stored in a database. The technique has been improved by other researchers [74, 76]. Image segmentation based on colour uses algorithms to process an image and extract coloured objects from the background for further analysis. It remains the most widely used colour-based approach. Several authors have reported techniques for colour segmentation: including clustering in colour space [114], region splitting [69, 88, 89], colour edge detection [64, 90], new parallel segmentation method based on 'region growing' or 'region collection' [93].

Shape-based sign detection relies largely on the significant achievements realised in the field of object recognition through research, such as techniques for scene analysis by robots, solid (3D) object recognition and part localisation in CAD databases. Almost all sign recognition systems process the colour information first to reduce the search for shape-based detection. Kehtarnavaz [81] extracted shapes from the image by performing edge detection and then applying the Hough transform to characterise the sides of the sign. Akatsuka [60] performed shape detection by template matching. de Saint-Blancard [68] used neural networks or expert systems as sign classifiers for a set of features consisting of perimeter (number of pixels), outside surrounding box, surfaces (inside/outside contour within surrounding box), centre of gravity, compactness ('aspect ratio' of box), polygonal approximation, Freeman code, histogram of Freeman code, and average grey level inside of box. Kellmeyer [82] trained a multi-layer neural net (with Back Propagation) to recognise diamond-shape warning signs in colour-segmented images. Piccioli [91, 92] concentrated exclusively on geometrical reasoning for sign detection, detecting triangular shapes with Canny's algorithm and circles in a Hough-like manner. On the other hand, Priese [94] worked on a model-based approach where basic shapes of traffic sign components (circles, triangles, etc.) are predefined with 24-edge polygons describing their convex hulls. Segmented images are first scanned for 'objects', which are then encoded and assigned a probability (based on an edge-to-edge comparison between the object and the model) for shape classification. Besserer [62] used knowledge sources (a corner detector, a circle detector and a histogram-based analyser) to classify chain coded objects into shape classes. Other techniques, referred to as 'rigid model fitting' in [110], have also been used for shape-based sign detection. Stein [109], Lamdan [84] and Hong [77] use specific model representations and a common matching mechanism, geometric hashing, to index a model database.

The PCNN technique developed here does not necessitate any colour or shape processing. The automatic identification of signs is achieved simply through matching of the barcodes of the images with the barcodes stored in the library (Fig. 7.28) [100].

An unknown sign can therefore be rapidly recognised using its unique barcode; a set of standard navigational signs is shown in Fig. 7.29 along with their respective barcodes.

7.4.6 Hand Gesture Recognition

Hand gesture recognition provides a natural and efficient communication link between humans and computers for human computer interaction and robotics [71, 78]. For example, new generations of intelligent robots can be taught how to handle objects in their environments by watching human subjects (if not other robots) manipulating them. Unlike most modes of communication, hand gestures usually possess multiple concurrent characteristics. Hand gestures can be either static, like a pose, or dynamic (over space and time) and include the hand gestures/hand signs commonly used in natural sign languages like the American Sign Language (ASL) or Australian Sign Language (AUSLAN). Although, there are many methods currently being exploited for recognition purposes, using both the static and dynamic characteristics of hand gestures, they are computationally time demanding, and therefore, not suitable for real-time applications. Recognition methods can be classified into two main groups, those requiring special gloves with sensors and those using computer vision techniques [66,85,116]. Recognition methods that fall under the first category can give very reliable information. Unfortunately, the connection cables in the gloves highly limit human movements in addition to being unsuitable for most real-world applications. Consequently, interests in computer vision techniques for hand gesture recognition have grown rapidly during the last few years.

Several researchers have devised hand gesture recognition systems in which marks are attached on fingertips, joints, and wrist [66]. Despite being suitable for real-time processing, it is however inconvenient for users. Another approach uses electromagnetic sensors and stereo-vision to locate the signer in video images [116]. To recognise ASL signs, Darrell [65] adopts a maximum a posteriori probability approach and uses 2D models to detect and tract human movements. Motion trajectories have also been utilised for signer localisation [72]. However, these approaches require a stationary background with a certain predetermined colour or restrict the signer to wear specialised gloves and markers, which makes them unsuitable for most real-world applications. Researchers have also investigated the use of neural network based systems for the recognition of hand gestures. These systems should enable major advances in the fields of robotics and human computers interaction (HCI). Using artificial neural systems, Littmann [86] demonstrate the visual recognition of human hand pointing gestures from stereo pairs of video camera images and provide a very intuitive kind of man-machine interface to guide robot movements. Based on Johansson's suggestion that human ges-

7.4 Barcode Applications

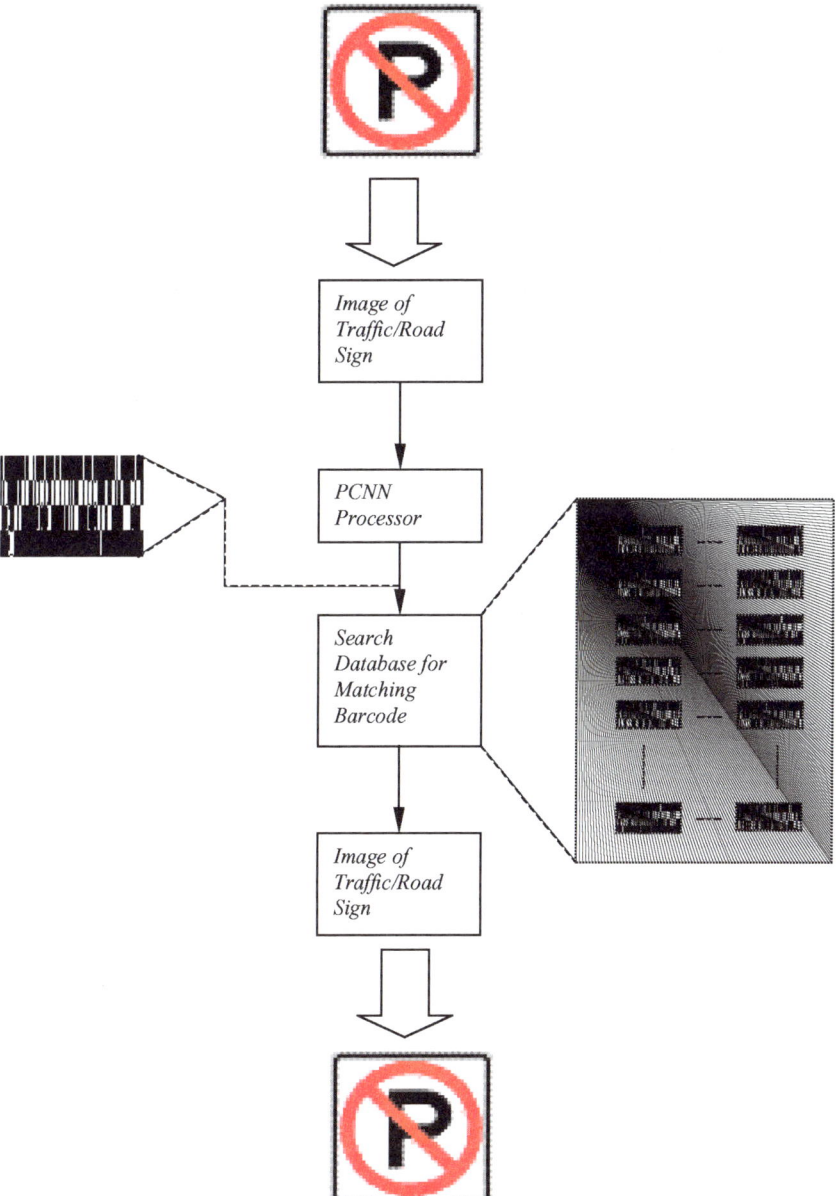

Fig. 7.28. Barcode generation from still road signs

136   7 Miscellaneous Applications

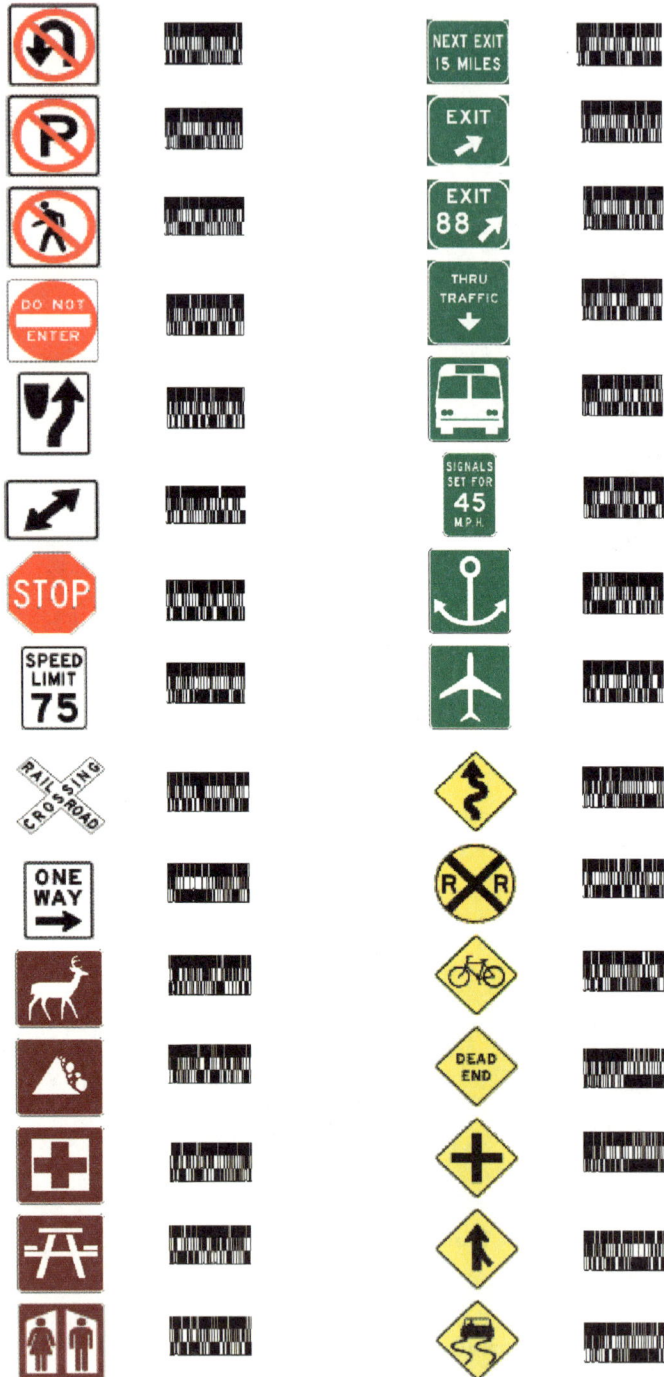

Fig. 7.29. Typical road signs and their corresponding barcodes

ture recognition rests solely on motion information, several researchers have carried out investigations on motion profiles and trajectories to recognise human motion [80]. Siskind [105] demonstrated gesture classification based on motion profiles using a mixture of colour based and motion based techniques for tracking. Isard [79] have come forward with the CONDENSATION algorithm as a probabilistic method to track curves in visual scenes. Furthermore, Yang [117] have used time-delay neural network (TDNN), specifically trained with standard error back propagation learning algorithm, to recognise hand gestures from motion patterns.

The one-to-one correspondence between each image and its corresponding binary barcode is shown in Fig. 7.30 [97]. Recognition of hand gestures is performed using an N-tuple weightless neural network.

7.4.7 Road Surface Inspection

Inspections of road surfaces for the assessment of road condition and for locating defects including cracks in road surfaces are the traditionally carried out manually. As such they are time-consuming, subjective, expensive, and can prove to be hazardous and disruptive to both the road inspectors and the circulating traffic users. What is ideally required would be a fully equipped automated inspecting vehicle capable of high precision location (to the nearest cm) and characterization of road surface defects (say cracks of widths 1 mm or greater) over the width of the road at speeds (up to $80\,\mathrm{kmh}^{-1}$) over a single pass. The automated system could also be enhanced to store the type of cracks present.

Several studies on automated systems for the detection and characterization of road cracks have been reported recently [75, 95, 113]. In this spirit, Transport Research Laboratory Ltd. (UK) as recently proposed an automatic crack monitoring system, HARRIS [95]. In this system video images of the road surface are collected by three linescan cameras mounted on a survey vehicle with the resolution of the digital images being 2 mm of road surface per pixel in the transverse direction and a survey width of 2.9 m. The scanned image (256 KB) is preprocessed to 64 KB (through reduction of gray levels). Reduced images are then stored in hard disk together with the location information. The location referencing subsystem reported in HARRIS (± 1 m accuracy) requires extra cameras and other hardware. The image processing of HARRIS is carried out in two stages: the first one consists of cleaning and reducing the images (on-line operation aboard the vehicle) and the second stage consists of an off-line operation on the reduced images to characterize the nature of the cracks. A typical one day survey of 300 km of traffic lane would tantamount to 80 GB of data collected. Full details of HARRIS can be found elsewhere citePynn99.

Several refinements to the HARRIS system for a more robust automatic inspection system are obvious. First, Global Positioning Systems (GPS) can be used (instead of video-based subsystem) to provide a much better accuracy

138   7 Miscellaneous Applications

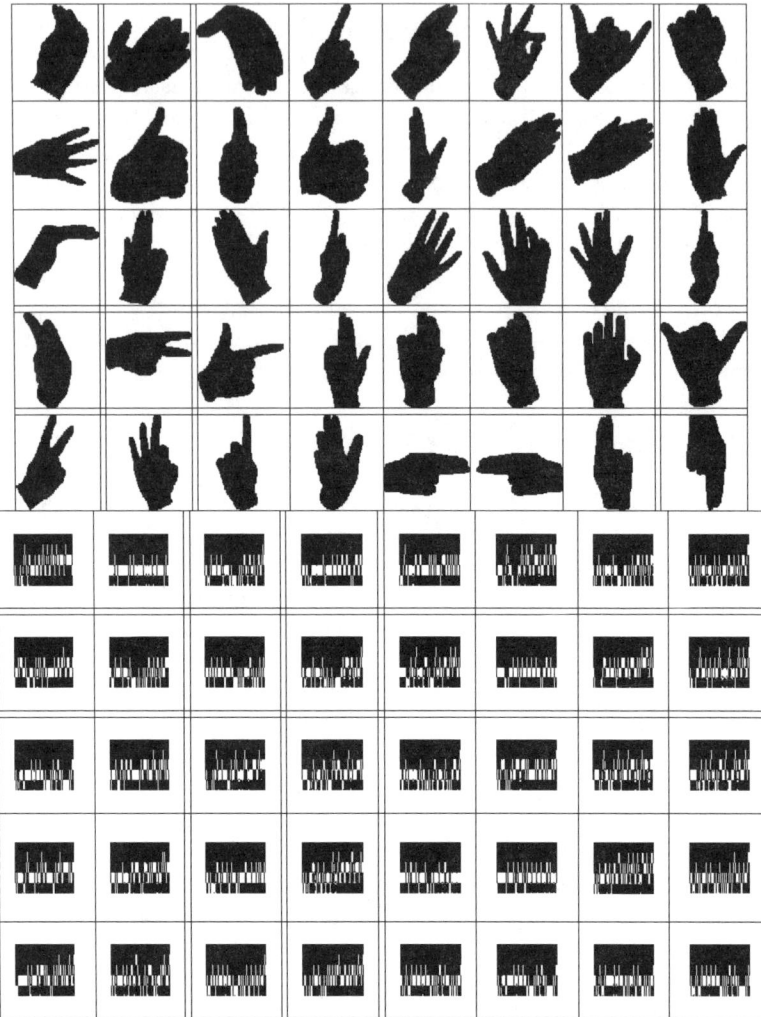

Fig. 7.30. 2D hand gestures used in experiments and their corresponding barcodes

for position location (down to 1 mm with differential GPS). Second, there is no need to store large volumes of scanned images of 'acceptable' road surface conditions. In this respect, the PCNN technique can be used for preprocessing each scanned image to detect defects and a second PCNN to segment this image if any defect(s) is (are) identified [Rughooputh,00b]. The latter image is then stored as binary image along with the GPS data. A real-time-crack map can be displayed by combining the results of the individual cameras. Detailed characterization of the defects can be performed offline from the recorded binary images. This mode of data collection leads to a more accurate, less costly and faster automated system.

Since the reflective responses of the material road surface can differ from place to place, there is a need to calibrate the software with a sample of a good road surface condition. This can be done in real-time in two modes – either once if the inspector is assured that for the whole length of the road has the same reflective responses in which case one sample image will suffice or periodically taking sample images at appropriate intervals. The overall philosophy behind the success of our method relies on the comparison of the input images from the camera to a reference (calibrated sample) image representing the good surface condition from defects. Our method does not compare the image directly since this will be very time-consuming; instead it involves the conversion of the input image into a binary barcode. The barcode generated from an image containing a defect will be different from that of a good road surface condition.

The basis of the crack identification process is the fact the barcode of an image containing a crack (whatever the nature) is different from the barcode representing a good surface condition. Figure 7.31a shows examples of different road conditions - the input images being stored in the PGM format (256 levels). The binary images with defects collected at the end of the surveyed road can then be analysed both in real time and off-line. Figure 7.31b shows the binary images of the segmented images using PCNN. It is clear that the nature of the defects depicted from these images can be easily classified (crack widths, hole sizes, etc.) according to an established priority for remedial actions. Since cracks can be observed as dark areas on the digital images, a second order crack identification algorithm can be used to crudely classify the priority basis. In this case, a proper threshold level needs to be found after carefully omitting redundant parts of the images (for e.g. corner shadowing effects).

When compared with the performance of HARRIS, the technique reported offers a much higher success rate (100%) for the crack identification process [98]. The high false-positive rates i.e. low success rate of HARRIS can be attributed to the poor image qualities arising from the fact that the number of grey levels in the cleaned images have been reduced in the primary processing stage (to 64 KB for storage purposes). Unlike HARRIS, there is no necessity to add specific criteria in our crack identification process, the need to add predetermined criteria to join crack fragments, and the need to store large volumes of scanned images of 'acceptable' road surface conditions. HARRIS also had to inbuilt a special algorithm based on predetermined criteria to join crack fragments. In short, we save a lot in terms of computing time. We note that road markings (such as yellow or white indicators) and artificially created structures (such as manholes and construction plates, etc), would be initially treated as defects. In any case, most of such markings occur on either side of the lane so that the camera can be adjusted to reveal say around 80–90% of the road width. In this configuration, most of the road markings will be

140   7 Miscellaneous Applications

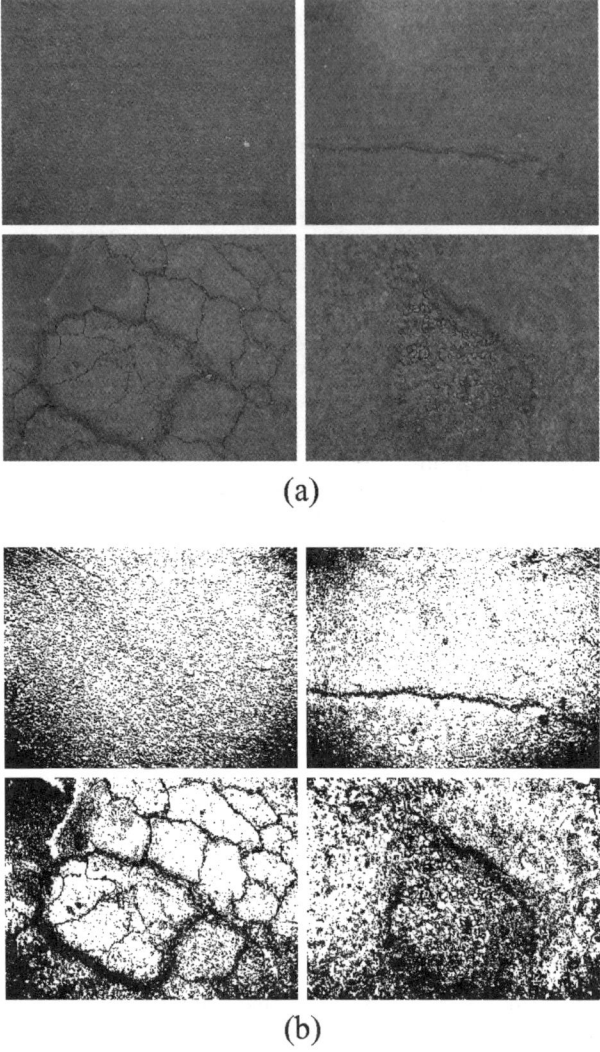

Fig. 7.31. (a) Typical road conditions (b) Segmented images of typical road conditions shown in (a)

ignored. Other markings or structures can be easily rejected when the binary images are analysed manually. Using the GPS location data, the collected binary segmented images from each camera are laid side by side in multi-lane surveys to create a crack map which indicates not only the location, length and direction of each crack identified, but also identifies cracks or defects extending beyond the boundaries of the individual survey cameras. We note that it is possible to obtain crack maps in real-time in single passes using our technique.

7.5 Summary

Obviously, there are a variety of applications that the PCNN or ICM can be of assistance. The main purpose of the PCNN family of algorithms is usually to isolate the important data for further analysis. However, there are also applications in which the PCNN was the main tool for condensing the original information into its final form. This chapter does not cover the full gambit of applications, but it does provide a taste of the wide variety of applications in which the PCNN is useful.

8 Hardware Implementations

While the PCNN is easy to employ in a standard computer algorithm, there have been a few implementations of the PCNN directly into hardware.

8.1 Theory of Hardware Implementation

To surpass the performance of a software implementation of the PCNN the hardware implementation usually relies on parallelization. Most of the operations in the PCNN are contained within the neurons, and the only real concern of making a parallel algorithm is the inter-neuron communications.

A hardware implementation can also produce similar PCNN results without employing the full algorithm. The ICM was developed after recognizing that there is a minimal set of calculations needed to produce results similar to the PCNN. The pursuit is to determine the communications necessary to create autowaves. In many simulations each neuron is allowed to communicate with several of its neighbours. In many of the biological models the neurons were connected to only their nearest neighbours

The images in Fig. 8.1 explore the possibility of communication with 1, 2 and 3 random neighbours. Figure 8.1a is the original input which would continually expand in this simulation. Figure 8.1b demonstrates the output of the same system in which the neurons communicate with a only single neighbour. As can be seen, the output did not demonstrate an expansion. Instead, the activity decayed and stabilized after 10 iterations.

Figure 8.1c demonstrates the same system with 2 neuron communications. Again this system would stabilize within a few iterations. Figure8.1d shows behaviour more similar to the original system. Here the boundaries of the original items expand in all directions, although not very smoothly. The system did not stabilize and the boundaries continued to expand. From this preliminary examine it appears that in the case of random connections at least three connections for each neuron are necessary to keep the boundary expanding. This is feasible in hardware constructs.

The minimal requirements for construction of these models are shown in Fig. 8.2. This neuron has an internal accumulator, U, and a threshold, Γ. When the neuron fires a feedback depletes the accumulation. This basic system is contained in the many models presented above.

144   8 Hardware Implementations

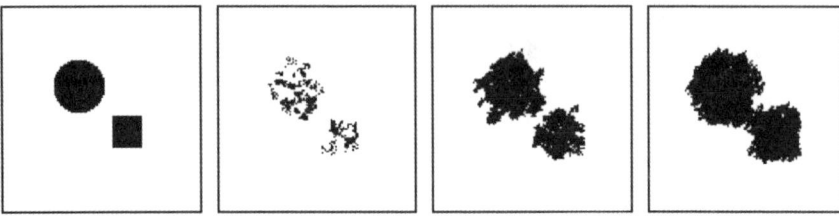

Fig. 8.1. Original input and outputs from 10 iterations for 1, 2, and 3 randomly selected connections

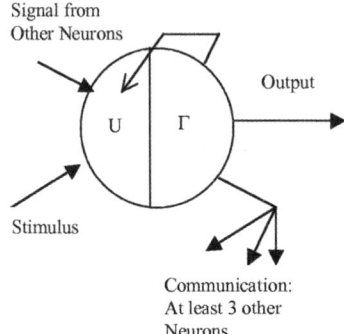

Fig. 8.2. Minimal cortical neuron

Hardware versions of these models may vary from the original model due to hardware limitations. However, if the hardware models are founded in the minimal system, the performance of the hardware model will produce similar results. To use these models to solve other problems (e.g., target recognition, object isolation, foveation, etc.) hardware constructions based on Fig. 8.2 will provide similar performance characteristics as the biological models. Thus, exact replication of the biological model is not necessary.

8.2 Implementation on a CNAPs Processor

Implementation of the PCNN in a parallel architecture can be fairly straightforward [123]. Unfortunately, the architectures of each parallel computer provide their own 'anomalies' and the example given here will therefore not necessarily be transferable to other architectures.

The example given involves a PC expansion slot card that uses a Single Instruction Multiple Data (SIMD) architecture named CNAPS manufactured by Adaptive Solutions Inc. This architecture consists of a number (P) of processing nodes or PNs. Each node has the ability to perform simple multiply and summing computations and has 4 K of local memory. In the SIMD architecture each node receives and executes the same instruction at the same

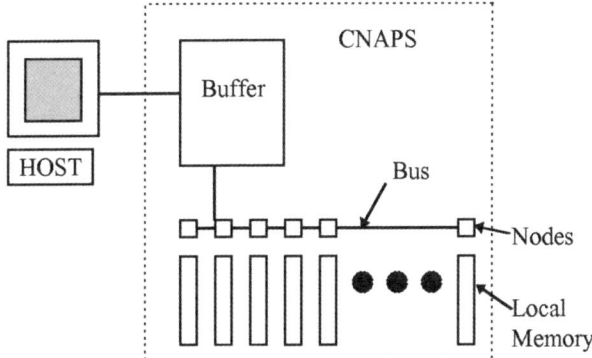

Fig. 8.3. The schematic of the CNAPS

time. Each node is restricted to operate on data within its local memory. Between the host computer and the CNAPS is a large buffer. This architecture is displayed in Fig. 8.3.

As mentioned above, the implementation of the PCNN in CNAPS is fairly straightforward. The only complication was the convolution and the distribution of the previous iteration's results for the convolution. The data can be transferred between nodes on a single bus. Unfortunately, the bus is a serial device, and therefore computations from one node using data from another node forces the algorithm into a serial mode, which destroys the whole point of parallel implementation.

Therefore, two modifications were made. The first is that the local memory in each node needs to store all of the values of K necessary for that node's computations. The computations of the convolution are stored in W. In the next iteration it is necessary to distribute the values of Y to the appropriate nodes for the convolution computation. This needs to be done in a parallel process on a serial bus. Fortunately, for a 5×5 kernel the Y values only need to travel one a or two nodes distance. Adaptive Solutions does provide commands that will shift data to neighbouring nodes simultaneously. Even though several values exist on the bus at the same time, they do not exist at the same place on the bus. Using these shift commands maintains the parallel distribution of the data.

The PCNN computation then looks as follows

```
/* This routine performs all of the PCNN functions for
a single iteration */

void Iterate( void )
{
   mono int i;
   Convolution( );
   [domain neuron].{
```

```
            foralli {
                F[i] = eaf * F[i] + S[i] + Vf * W[i];
                L[i] = eal * L[i] + Vl * W[i];
                U[i] = F[i] * ( 1.0 + beta * L[i] );
                T[i] = eat * T[i] + Vt * Y[i];
                if( U[i] > T[i] + 0.1 ) Y[i] = 1;
                else Y[i] = 0;
            }
        }
    }
```

Each array variable is replicated on each node so each node only has to perform N operations instead of $N \times N$. Once the convolution is performed the computations are quite straightforward.

On a 90 MHz Pentium 20 iterations took about 25 seconds, and on the CNAPS the same 20 iterations took about a second.

8.3 Implementation in VLSI

There are several ways to implement the PCNN in silicon. It may not always be desirable to have a fixed configuration with relatively little control from outside. On the other hand, it may sometimes be quite useful to have the direct pre-processing provided by the PCNN implemented immediately after the sensor. Intelligent sensors have been discussed for a long time and there are image arrays with integrated processors like the LAPP 1100, 1510 and MAPP2200 from IVP [119, 122].

The LAPP 1100 is a line sensor with 128 pixels (24 microns) and 128 parallel processing elements (PE) with a complexity of 320 transistors per PE. This device may thus be characterised as a detector with an integrated processor for image processing applications. The MAPP2200 is a 256×256 pixels 2D sensor containing a 1D SIMD processor with a complexity of 1500 transistors per PE. A 2D/2D solution called NISP [120] has recently been developed with $32*32$ PE and where each processor only has 110 transistors.

Another photodiode device has been suggested by Guest [121] and is shown in Fig. 8.4. It employs a very simple PCNN and has still fewer transistors.

8.4 Implementation in FPGA

Field programmable logic devices, in particular Field Programmable Gate Arrays (FPGAs), are without doubt one of the most important technologies of recent years. In particular, they revolutionized system design and now offer very quick turn-around. In particular large RAM available on the FPGAs

8.4 Implementation in FPGA   147

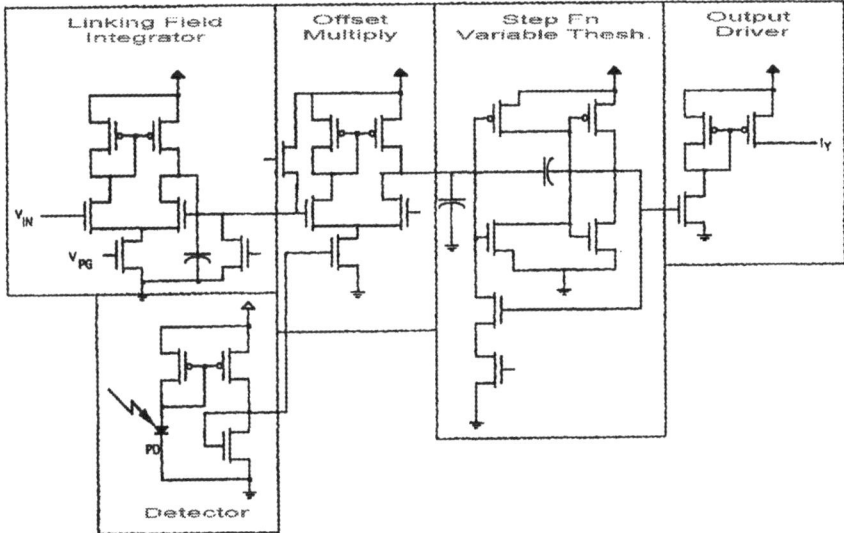

Fig. 8.4. Implementation of a simple PCNN directly following a photodiode

encouraged rapid experimentation. At the same time, a lot of commercial tools and libraries become available. Software for design, analysis and synthesis become very advanced in all respects and the VHDL is commonly used. The tools will change rapidly particularly as the mainstream designs employ FPGAs as primary logic source, and not just as ASIC prototype vehicles.

The VHDL code below [124] describes one single PCNN neuron that has one image intensity input (8 bit grey scale) and 8 feedback inputs from surrounding neurons. The PCNN neuron needs many multiplication steps for each iteration: especially the merging operator of the linking and feeding branches particularly is critical. These multiplications are implemented as 22×22 ones in this simple example below.

```
library ieee;
use ieee.std_logic_1164.all;
use ieee.std_logic_arith.all;
package pcnn_package is
   constant beta_VL_width:natural:=3+15; -- [0..1]*[0..10]
   constant beta_VL_binal:natural:=15; -- res 0.001*0.05
   constant Vf_width:natural:=1+11; -- [0..1]
   constant Vf_binal:natural:=11; -- res 0.0005
   constant Vt_width:natural:=6+4; -- [1..50]
   constant Vt_binal:natural:=4; -- res 0.1
   constant exp_width:natural:=1+15; -- [0..1]
   constant exp_binal:natural:=15; -- res 5E-5
```

```vhdl
   constant beta_VL:unsigned(beta_VL_width-1 downto 0);
   constant Vf:unsigned(Vf_width-1 downto 0);
   constant Vt:unsigned(Vt_width-1 downto 0);
   constant KL:unsigned(exp_width-1 downto 0);
   constant KF:unsigned (exp_width-1 downto 0);
   constant alfa_T:unsigned(exp_width-1 downto 0);

end pcnn_package;

package body pcnn_package is
   constant beta_VL:unsigned(beta_VL_width-1 downto 0):=
   conv_unsigned(integer(0.01*0.5*2**beta_Vl_binal),
       beta_VL_width);
   constant Vf:unsigned(Vf_width-1 downto 0):=
   conv_unsigned(integer(0.03*2**Vf_binal),Vf_width);
   constant Vt:unsigned(Vt_width-1 downto 0):=
   conv_unsigned(integer(39.1*2**Vt_binal),Vt_width);
   constant KL:unsigned(exp_width-1 downto 0):=
   conv_unsigned(integer(0.36*2**exp_binal),exp_width);
   constant KF:unsigned (exp_width-1 downto 0):=
   conv_unsigned(integer(0.25*2**exp_binal),exp_width);
   constant alfa_T:unsigned(exp_width-1 downto 0):=
   conv_unsigned(integer(0.16*2**exp_binal),exp_width);
end pcnn_package;

library ieee;
use ieee.std_logic_1164.all;
use ieee.std_logic_arith.all;
use work.pcnn_package.all;
entity pcnn is
   port(clk:IN std_logic;
      reset:IN std_logic;
      Y0,Y1,Y2,Y3,Y4,Y5,Y6,Y7:IN unsigned(0 downto 0);
      S:IN unsigned(7 downto 0);
      Y:INOUT unsigned(0 downto 0));
end pcnn;

architecture behave of pcnn is

   signal sum:unsigned(3 downto 0);

   signal Linking:unsigned(4+beta_VL_width-1 downto 0);
   signal L,L_reg:unsigned(4+beta_VL_width-1 downto 0);
```

8.4 Implementation in FPGA

```vhdl
   signal L_mult_KL:unsigned(4+beta_VL_width+exp_width-1
       downto 0);
     -- L_mult_KL_binal equals exp_binal+beta_VL_binal
     -- The signal should be added to Linking, which equals
        beta_VL_binal
     -- Thus, the final signal equals beta_VL_binal and
        exp_binal is dropped

   signal L_one:unsigned(4+beta_VL_width-1 downto 0);

   signal Feeding:unsigned(4+Vf_width-1 downto 0);
   signal F,F_reg:unsigned(8+Vf_width-1 downto 0); --
       128 iterations + Y
firing
   signal F_mult_KF:unsigned(8+Vf_width+exp_width-1
       downto 0);
     -- F_mult_KF equals exp_binal+Vf_binal
     -- The signal should be added to feeding, which equals
        Vf_binal
     -- Thus, the final signal equals Vf_binal and
        exp_binal is dropped

   constant F_zero:unsigned(Vf_binal-8-1 downto 0):
       =(others=>'0');

   signal L_mult_F:unsigned(8+Vf_width+4+beta_VL_width-1
       downto 0);
     -- Should actually be +2*(exp_width-exp_binal) more
        bits, but exp_width
and
     -- exp_binal should be the same
     -- The signal should be compared to theta, which
        equals Vt_binal
     -- Thus, the final signal equals Vt_binal and
        Vf_binal+beta_VL_binal-Vt_binal
     -- is dropped

   signal U:unsigned(8+Vf_width+4+beta_VL_width-1 downto
       Vf_binal+beta_VL_binal-Vt_binal);

   signal theta,theta_reg:unsigned(Vt_width-1 downto 0);
   signal theta_mult_alfa_t:unsigned(Vt_width+exp_width-1
       downto 0);
     -- theta_mult_alfa_t equals Vt_binal+exp_binal
     -- The signal should be compared to U, which
```

                equals Vt_binal
             -- Thus, the final signal equals Vt_binal and
                exp_binal is dropped

   begin

   sum<=(("000"&Y0)+("000"&Y1)+("000"&Y2)+("000"&Y3))+
        (("000"&Y4)+("000"&Y5)+("000"&Y6)+("000"&Y7));

      Linking<=sum*beta_VL;
      Feeding<=sum*Vf;

      L_mult_KL<=L_reg*KL;
      F_mult_KF<=F_reg*KF;

      L<=Linking+L_mult_KL(4+beta_VL_width+exp_width-1 downto
           exp_binal+1);
      F<=Feeding+F_mult_KF(8+Vf_width+exp_width-1 downto
           exp_binal+1)+(S & F_zero);

      L1:for i in 4+beta_VL_width-1 downto 0 generate
           L_one(i)<='1' when i=beta_VL_binal else '0';
         end generate;

      L_mult_F<=(L_one+L)*F;

      U<=L_mult_F(8+Vf_width+4+beta_VL_width-1 downto
           Vf_binal+beta_VL_binal-Vt_binal);

      Y<=unsigned'("1") when U>theta else unsigned'("0");

      theta_mult_alfa_t<=theta_reg*alfa_t;
      theta<=theta_mult_alfa_t(Vt_width-1 downto 0);
      process(clk)
      begin
           if (clk'event and (clk='1')) then
         if (reset='1') then
           L_reg<=(others=>'0');
           F_reg<=(others=>'0');
         else
           L_reg<=L;
           F_reg<=F;
         end if;
           end if;
      end process;

```
    process(clk)
    begin
        if (clk'event and (clk='1')) then
      if ((Y=unsigned'("1")) OR (reset='1')) then
          theta_reg<=work.pcnn_package.Vt;
      else
          theta_reg<=theta;
      end if;
        end if;
    end process;
```

end behave;

The PCNN neuron above uses almost 1150 logic cells of a FPGA, which is about 22% of an ALTERA FLEX 10K100 chip. The neuron can process input/output in little over 4 MHz. The implementation is, however, not a very efficient one. If we assume a 128 × 128 pixel image having 8 bit resolution and a 60–70 Hz image speed we will get 16 ms total calculation time available for each image. For each image we need at least 100 PCNN iterations (that is, total net iterations on the input image), which gives us 0.16 ms per network iteration. If we solve this serially, then we have only 0.16/(128∗128) ms/neuron, that is almost 10 ns/neuron. Thus this Altera chip cannot process an image in real time using serial mode.

If we pipeline the critical multiplication steps of the PCNN neuron and assume that we can fit 4 neurons on a 10K100 or 10K130 chip, we could process approximately 64 pixels within 4 µs. This means one complete net iteration within 1 ms. However, this is still too slow for real time purpose using the Altera Flex family. However, the Altera 10K250 chip would fit about twice the number of neurons, giving a total network iteration time of about 0.5 ms. That is the 10K250 will be able to process 128 × 128 pixel resolution in close to 20 Hz speed without modifications of the suggested VHDL code.

The VHDL code above can be modified to be more optimal with respect to available resources. Thus the multiplication can be moved from the chips gate structure into the embedded array block (EAB) area on the chip, and in the same time speed up the process.

8.5 An Optical Implementation

Johnson [7] built an optical system that performed PCNN-like computations. This system was elegant in its simplicity and produced results similar to those expected from a PCNN. The schematic of this system is shown in Fig. 8.5.

This system illuminated an object with a diffuse white light source. The radiation from this source was focused on a plane that was slightly off the

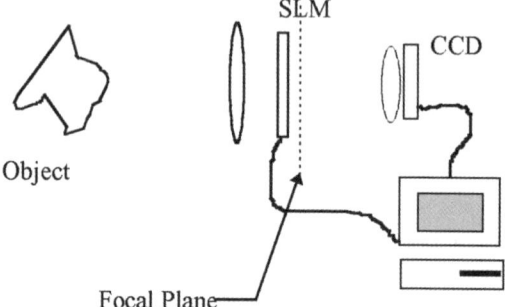

Fig. 8.5. An optical implementation of PCNN-like computations

Spatial Light Modulator (SLM) plane. A CCD detector received the image and fed it to the computer, which in turn, wrote an image onto the SLM.

The SLM was a transmissive device whose elements could be in either of two states. Both states transmitted light but the OFF state transmitted significantly less light than the ON state. The image passed through the SLM in a slightly out-of-focus state. As it passed through the SLM the image was multiplied by the image on the SLM but the off focus nature also performed the local interconnections. More exactly, this passing of the image produced a local convolution of the elements of the image with the elements of the SLM.

Let the input image be S and the SLM image be A. The image at the focal plane is F,

$$F_{ij} = S_{ij} \sum_{kl} m_{ijkl} A_{kl}, \qquad (8.1)$$

where m indicates the local interconnections produced by the off focus nature of the image.

The CCD detects the energy of the incoming image so the elements F_{ij}^2 are read from the CCD. It should be noted that F_{ij} are positive and incoherent (no phase) so the detection process does not significantly alter the data.

The data is then sent to the computer for a threshold operation. Thresholds are very difficult to implement optically so this step is performed in the computer. Basically, the computer performed the operation,

$$A_{ij} = \begin{cases} 1 & \text{if } |F_{ij}|^2 > \gamma \\ 0 & \text{Otherwise} \end{cases}, \qquad (8.2)$$

where the ON and OFF states of the SLM are indicated by 1 and 0, respectively, γ is a constant that is dependent upon overall illumination, detection bias, etc.

As can be seen, the optical system performed operations different in their mathematical form than the PCNN. However, the output images were similar to those of the PCNN output. Thus, in practice, this optical system could be used to simulate the PCNN.

The main advantage of using an optical system is that it can perform all interconnections fully in parallel. For large convolution kernels this can be an advantage. The speed of the system is limited by the speed of the SLM and CCD. Common SLMs and CCDs operate at 30 Hz, but newer devices with much higher speeds are being developed.

8.6 Summary

A variety of methods have been explored for implementing the PCNN or equivalent into a hardware architecture. The purpose of this is to provide the fastest speed possible for computing the pulse images. However, many of these architectures were developed back when desktop computers were running below 100 MHz. As computers have become faster by more than a factor of 10 since that time the need for hardware implementations is lessening. This bodes well for the PCNN and ICM as it becomes easier to employ them without specialized hardware.

References

1. R. Eckhorn, H. J. Reitboeck, M. Arndt, P. Dicke: Feature linking via synchronization among distributed assemblies: Simulations of results from Cat Visual Cortex. Neural Comp. **2**, 293–307 (1990)
2. U. Ekblad: Earth satellites and air and ground-based activities. Thesis, Royal Institute of Technology, Department of Physics, Trita-FYS. 2002:42
3. U. Ekblad, J.M. Kinser: Theoretical foundation of the intersecting cortical model and its use for change detection of aircraft, cars and nuclear explosion tests. Signal Processing **84**, 1131–1146 (2004)
4. U. Ekblad, J.M. Kinser, J. Atmer, N. Zetterlund: The intersecting cortial model in image processing. Nucl. Instr. Meth. A **525**, 392–396 (2004)
5. R. FitzHugh: Impulses and phsyiological states in theoretical models of nerve membrane. Biophysics J. **1**, 445–466 (1961)
6. A.L. Hodgkin, A.F. Huxley: A quantitative description of membrane current and its application to conduction and excitation in nerve. Journal of Physiology **117**, 500–544 (1952)
7. J.L. Johnson: Pulse-Coupled Neural Nets: Translation, rotation, scale, distortion, and intensity signal invariances for images. Appl. Opt. **33** (26), 6239–6253 (1994)
8. J.M. Kinser: The determination of hidden neurons. Optical Memories and Neural Networks **5** (4), 245–262 (1996)
9. A. Labbi, R. Milanese, H. Bosch: A network of FitzHugh–Nagumo oscillators for object segmentation. Proc. of International Symposium on Nonlinear Theory and Applications, NOLTA'97, Nov. 29–Dec. 3, Hawaii 1997, pp. 581—584
10. J. Nagumo, S. Arimoto, S. Yoshizawa: An active pulse transmission line stimulating nerve axon. Proc. IRE **50**, 2061–2070 (1962)
11. O. Parodi, P. Combe, J.-C. Ducom: Temporal encoding in vision: Coding by spike arrival times leads to oscillations in the case of moving targets. Biol. Cybern. **74**, 497–509 (1996)
12. I.A. Rybak, N.A. Shevtsova, V.A. Sandler: The model of a Neural Network visual processor. Neurocomputing **4**, 93-102 (1992)
13. Y.I. Balkarey, M.G. Evtikhov, M.I. Elinson: Autowave media and Neural Networks. SPIE **1621**, 238–249 (1991)
14. W. Gernster: Time structure of the activity in Neural Network Models. Phys. Rev. E **51** (1), 738–758 (1995)
15. M.A. Grayson: The heat equation shrinks embedded plane curves to round points. J. Differential Geometry **26**, 285–314 (1987)

16. H.S. Ranganath, G. Kuntimad: Image segmentation using pulse coupled neural networks. IEEE World Congress on Computational Intelligence, 1994 IEEE International Conference on Neural Networks, 1994, vol 2, pp. 1285–1290
17. J.L. Johnson, M.L. Padgett: PCNN models and applications. IEEE Trans. on Neural Networks, vol. 10, issue 3, May 1999 (Guest editorial overview of Pulse Coupled Neural Network (PCNN), special issue: Johnson, J.L.; Padgett, M.L.; Omidvar, O., pp. 461–463, 480–498)
18. J.M. Kinser: Hardware: Basic requirements for implementation. Proc. of SPIE **3728**, Stockholm, June 1998, 222–229
19. J.M. Kinser: Image signatures: Classification and ontology. Proc. of the 4th IASTED Int. Conf. on Computer Graphics and Imaging, 2001
20. R. Malladi, J.A. Sethian: Level set methods for curvature flow, image enhancement, and shape recovery in medical images. Proc. of Conf. on Visualization and Mathematics, June 1995 (Springer 1995) 329–345
21. C. McEniry, J.L. Johnson: Methods for image segmentation using a Pulse-Coupled Neural Network. Neural Network World 2/97, 177–189 (1997)
22. R.E. Mirollo, S.H. Strogatz: Synchronization of pulse-coupled biological oscillators. SIAM J. of Appl. Math. **50** (6), 1645–1662 (1990)
23. O.A. Mornev: Elements of the optics of autowaves. In: V.I. Krirsky (Ed.): *Self-Organization Autowaves and Structures far from Equilibrium* (Springer-Verlag 1984) 111–118
24. E. Neibur, F. Wörgötter: Circular inhibition: A new concept in long-range interaction in the Mammalian Visual Cortex. Proc. IJCNN, vol. II, San Diego 1990, 367–372
25. M. Akay: Wavelet application in medicine. Spectrum, 50–56 (May 1997)
26. J. Brasher, J.M. Kinser: Fractional-power synthetic discriminant functions. Pattern Recognition **27** (4), 577–585 (1994)
27. J.L. Horner: Metrics for assessing pattern recognition. Appl. Opt. **31** (2), 165–166 (1992)
28. J.L. Johnson, M.L. Padgett, W.A. Friday: Multiscale image factorisation. Proc. Int. Conf. on Neural Networks, ICNN97, Houston TX, June 1997, Invited paper, 1465–1468
29. J.M. Kinser, J.L. Johnson: Stabilized input with a feedback Pulse-Coupled Neural Network. Opt. Eng. **35** (8), 2158–2161 (1996)
30. J.M. Kinser, T. Lindblad: Detection of microcalcifications by cortial stimulation. In: A.B. Bulsari and S. Kallio (Eds.) *Neural Networks in Engineering Systems* (Turku 1997) pp. 203–206, EANN'97, Stockholm June 1997
31. B.V.K.V. Kumar: Tutorial survey of composite filter designs for optical correlators. Appl. Opt. **31** (23), 4773–4801 (1992)
32. J. Moody, C.J. Darken: Fast learning in networks of locally tuned processing units. Neural Computation **1**, 281–294 (1989)
33. M.L. Padgett, J.L. Johnson: Pulse-Coupled Neural Networks (PCNN) and wavelets: Biosensor applications. Proc. Int. Conf. on Neural Networks, ICNN97, Houston TX, June 1997, Invited paper, 2507–2512
34. J. Waldemark, V. Bečanović, T. Lindblad, C.S. Lindsey: Hybrid Neural Networks for automatic target recognition, IEEE Conf. on System, Man and Cybernetics, SMC97, vol 4. pp. 4016–4021, Orlando, FL, USA, October 1997
35. G. Wilensky, N. Manukian: The projection Neural Network. Int. Joint Conf. on Neural Networks, vol. II, 1992, pp. 358–367

References

36. J. Waldemark, V. Bečcanović, U. Brännström, C. Holmström, M. Larsson, Th. Lindblad, C.S. Lindsey, Å. Steen: A Pulse-Coupled Neural Network pre processing of aurora images. In: A.B. Bulsari and S. Kallio (Eds.) *Neural Networks in Engineering Systems*, (Turku 1997) pp. 29–32, EANN'97, Stockholm June 1997
37. Å.J. Eide, J. Waldemark, V. Bečcanović, U. Brännström, C. Holmström, M. Larsson, I.M. Lillesand, Th. Lindblad, C.S. Lindsey, Å. Steen: A Pulse-Coupled Neural Network pre processing of aurora images. Proc. 2nd Workshop on AI Applications in Solar-Terrestrial Physics, July 29–31, 1997, Lund, Sweden, ESA WPP-148
38. L.-J. Cheng, T.-H. Chao, G. Reyes: Acousto-optic tunable filter multispectral imaging system. AIAA Space Programs and Technologies Conference, paper no. 92-1439, March 24–27, 1992
39. L.-J. Cheng, T.-H. Chao, M. Dowdy, C. LaBaw, C. Mahoney, G. Reyes, K. Bergman: Multispectral imaging systems using acousto-optic tunable filter. Infrared and Millimeter Wave Engineering, SPIE Proc. **1874**, 224 (1993)
40. J.M. Kinser: Object isolation. Optical Memories and Neural Networks **5** (3), 137–145 (1996)
41. J.M. Kinser: Object Isolation Using a Pulse-Coupled Neural Network. Proc. SPIE **2824**, 70–77, (1996)
42. J.M. Kinser: Pulse-coupled image fusion. Opt. Eng. **36** (3), 737–742 (1997)
43. J.M. Kinser, C.L. Wyman, B.L. Kerstiens: Spiral image fusion: A 30 parallel channel case. Optical Eng. **37** (02), 492–498 (1998)
44. H. Ranganath, G. Kuntimad, J.L. Johnson: Image segmentation using Pulse-Coupled Neural Networks. Proc. of IEEE Southeastcon, Rayleigh, N. C., 1995, 49–53
45. Y.Q. Chen, M.S. Nixon, D.W. Thomas: Statistical geometrical features for texture classification. Pattern Recognition **28** (4), 537–552 (1995)
46. Y.Q. Chen: Novel Techniques for image texture classification. PhD Thesis, University of Southampton, Department of Electronics and Computer Science, 1996
47. R.M. Haralick, K. Shanmugam, I. Dinstein: Textural features for image classification. IEEE Trans. on System, Man. Cybernetics **3**, 610–621 (1973)
48. J.F. Haddon, J.F. Boyee: Co-occurrence matrices for image analysis. IEEE Electronics and Communications Engineering Journal **5** (2), 71–83 (1993)
49. D.C. He, L. Wang: Texture features based on texture spectrum. Pattern Recognition **25** (3), 391–399 (1991)
50. J.M. Kinser: Fast analog associative memory. Proc. SPIE **2568**, 290–293 (1995)
51. K.I. Laws: Textured image segmentation. PhD Thesis, University of Southern California, Electrical Engineering, January 1980
52. http://www.cssip.elec.uq.edu.au/~guy/meastex/meastex.html
53. W.K. Pratt: *Digital Image Processing* (A Wiley-Interscience Publication 2001)
54. S. Singh, M. Singh: Texture analysis experiments with Meastex and Vistex benchmarks. In: S. Singh, N. Murshed and W. Kropatsch (Eds.): Proc. Int. Conf. on Advances in Pattern Recognition, Rio (11–14 March 2001), Lecture Notes in Computer Science **2013**, 417–424 (Springer-Verlag)

References

55. M. Singh, S. Singh: Spatial texture analysis: A comparative study. Proc. 15th Int. Conf. on Pattern Recognition (ICPR'02), Quebec, (11–15 August 2002)
56. M. Tuceryan, A.K. Jain: Texture analysis. In: *Handbook of Pattern Recognition and Computer Vision*, ed by C.H. Chen, L.F. Pau and P.S. Wang (World Scientific Publishing 1993) 235–276
57. M.R. Vasquez, P. Katiyar: Texture classification using logical operations. IEEE Trans. on Image Analysis **9** (10), 1693–1703 (2000)
58. J.M. Kinser, C. Nguyen: Image object signatures from centripetal autowaves. Pattern Recognition Letters **21** (3), 221–225 (2000)
59. J.W. McClurken, J.A. Zarbock, L.M. Optican: Temporal codes for colors, patterns and memories. Cerebral Cortex **10**, 443–467 (1994)
60. H. Akatsuka, S. Imai: Road signposts recognition system. Proc. SAE Vehicle Highway Infrastructure: Safety Compatibility, 189–196 (1987)
61. J. Arens, A. Saremi, C. Simmons: Color recognition of retroreflective traffic signs under various lighting conditions. Public Roads **55**, 1–7 (1991)
62. B. Besserer, S. Estable, B. Ulmer: Multiple knowledge sources and evidential reasoning for shape recognition. Proc. IEEE 4th Conference on Computer Vision, 624–631 (1993)
63. R. Buta, S. Mitra, G. de Vaucouleurs, H. G. Corwin: Mean morphological types of bright galaxies. Atronomical Journal **107**, 118 (1994)
64. T. Carron, P. Lambert: Color edge detector using jointly hue, saturation and intensity. IEEE Int. Conf. on Image Processing **3**, 977–981 (1994)
65. T. Darrell, I. Essa, A. Pentland: Task-specific gesture analysis in real-time using interpolated views. IEEE Trans. Pattern Anal. and Mach. Intell. **18** (12), 1236–1242 (1996)
66. J. Davis, M. Shah: Recognizing hand gestures. ECCV'94, 331–340 (1994)
67. G. de Vaucouleurs, A. de Vaucouleurs, H.G. Corwin, R. Buta, G. Paturel, P. Fouque: *Third Reference Catalog of Bright Galaxies* (Springer-Verlag NY 1991)
68. M. de Saint Blancard: Road sign recognition: A study of vision-based decision making for road environment recognition. In: I. Masaki (Ed.) *Vision-Based Vehicle Guidance* (Springer-Verlag, New York, Berlin, Heidelberg 1992) 162–172
69. M.P. Dubuisson, A. Jain: Object contour extraction using color and motion. IEEE Int. Conf. Image Processing, 471-476 (1994)
70. O.J. Eggen, D. Lynden-Bell, A.R. Sandage: Evidence from the motion of old stars that the galaxy collapsed. Astrophysical Journal **136**, 748 (1962)
71. S.S. Fels, G.E. Hinton: Glove-talk: A neural network interface between a data-glove and a speech synthesizer. IEEE Trans. Neural Network **4**, 2–8 (1993)
72. W.T. Freeman, C.D. Weissman: Television control by hand gestures. Proc. Int. Workshop on Automatic Face and Gesture Recognition, 179–183 (1995)
73. Z. Frei, P. Guhathakurta, J.E. Gunn, J.A. Tyson: A catalog of digital images of 113 nearby galaxies. Astronomical Journal **111**, 174–181 (1996)
74. B.V. Funt, G.D. Finlayson: Color constant color indexing. IEEE Trans. on Patt. Anal. Mach. Intell. **17** (5), 522–529 (1955)
75. L. Hawker: The introduction of economic assessment to pavement maintenance management decisions on the United Kingdom using private finance. XIIIth IRF World Meeting, Toronto, Ontario, Canada (1997)

76. G. Healey, D. Slater: Global color constancy: recognition of objects by use of illumination-invariant properties of color distributions. J. Opt. Soc. Am. A **11** (11), 3003–3010 (1994)
77. J. Hong, H. Wolfson: An improved model-based matching method using footprints. Proc. 9th Int. Conf. Pattern Recognition, IEEE, 72–78 (1988)
78. T.S. Huang, V.I. Pavlovic: Hand modelling, analysis, and synthesis. Int. Workshop on Automatic Face and Gesture Recognition, Zurich, June 26–28 (1995) pp. 73–79
79. M. Isard, A. Blake: Condensation – conditional density propagation for visual tracking. International Journal of Computer Vision **29** (1), 5–28 (1998)
80. G. Johansson: Visual perception of biological motion and a model for its analysis. Perception and Psychophysics **73** (2), 201–211 (1973)
81. N. Kehtarnavaz, N.C. Griswold, D.S. Kang: Stop-sign recognition based on color-shape processing. Machine Vision and Applications **6**, 206–208 (1993)
82. D. Kellmeyer, H. Zwahlen: Detection of highway warning signs in natural video images using color image processing and neural networks. IEEE Proc. Int. Conf. Neural Net **7**, 4226–4231 (1994)
83. D. Krumbiegel, K.F. Kraiss, S. Schreiber: A connectionist traffic sign recognition system for onboard driver information. 5th IFAC/IFIP/IFORS/IEA Symposium on Analysis, Design and Evaluation of Man-Machine Systems, 201–206 (1993)
84. Y. Lamdan, H. Wolfson: Geometric hashing: a general and efficient model-based recognition scheme. Proc. 2nd Int. Conf. on Computer Vision, IEEE, 238–249 (1988)
85. J. Lee, T.L. Kunii: Model-based analysis of hand posture. IEEE Computer Graphics and Applications, 77–86 (1995)
86. E. Littmann, A. Drees, H. Ritter: Visual gesture-based robot guidance with a modular neural system. Advances in Neural Information Processing Systems 8 (Morgan Kaufman Publishers, San Mateo, CA 1996) 903–909; E. Littmann, A. Drees, H. Ritter: Neural system recognizes human pointing gestures in real images. In: *Neuronale Netze in Ingenieursanwendungen* (ISD, Universität Stuttgart 1996) 53–64
87. M A. Miller: Chemical database techniques in drug discovery. Nature, 220–227 (2002)
88. R. Ohlander, K. Price, D. Reddy: Picture segmentation using a recursive region splitting method. Computer Graphics and Image Processing **8**, 313–333 (1978)
89. Y. Ohta, T. Kanade, T. Sakai: Color information for region segmentation. Computer Graphics and Image Processing **13**, 224–241 (1980)
90. F. Perez, C. Koch: Toward color image segmentation in analog VLSI: Algorithm and hardware Int. J. of Computer Vision **12** (1), 17–42 (1994)
91. G. Piccioli, E.D. Michelli, M. Campani: A robust method for road sign detection and recognition. Proc. European Conf. on Computer Vision, 495–500 (1994)
92. G. Piccioli, E.D. Michelli, P. Parodi, M. Campani: Robust road sign detection and recognition from image sequence. Proc. Intelligent Vehicles '94, 278–283 (1994)
93. L. Priese, V. Rehrmann: On hierarchical color segmentation and applications. Proc. CVPR, 633–634 (1993)

94. L. Priese, J. Klieber, R. Lakmann, V. Rehrmann, R. Schian: New results on traffic sign recognition. IEEE Proc. Intelligent Vehicles'94 Symposium, 249–253 (1994)
95. J. Pynn, A. Wright, R. Lodge: Automatic identification of road cracks in road surfaces. Proc. 7th Int. Conf. on Image Processing and Its Applications 2, Manchester (UK), 671–675 (1999)
96. M.S. Roberts, M.P. Haynes: Physical parameters along the Hubble sequence. Annual Review Astronomy and Astrophysics **32**, 115 (1994)
97. H.C.S. Rughooputh, H. Bootun, S.D.D.V. Rughooputh: Intelligent hand gesture recognition for human computer interaction and robotics. Proc. RESQUA2000: The First Regional Symposium on Quality and Automation: Quality and Automation Systems for Advanced Organizations in the Information Age, IEE, Universiti Sains, Malaysia, 346–352 (2000)
98. H.C.S. Rughooputh, S.D.D.V. Rughooputh, J. Kinser: Automatic inspection of road surfaces. In: K.W. Tobin, Jr (Ed.) *Machine Vision Applications in Industrial Inspection VIII*, Proc. SPIE **3966**, 349–356 (2000)
99. S.D.D.V Rughooputh, R. Somanah, H.C.S. Rughooputh: Classification of optical galaxies using a PCNN. In: N. Nasrabadi (Ed.) *Applications of Artificial Neural Networks in Image Processing V*, Proc. SPIE **3962** (15), 138–147 (2000)
100. S.D.D.V. Rughooputh, H. Bootun, H.C.S. Rughooputh: Intelligent traffic and road sign recognition for automated vehicles. Proc. RESQUA2000: The First Regional Symposium on Quality and Automation: Quality and Automation Systems for Advanced Organizations in the Information Age, IEE, Universiti Sains, Malaysia, May 4–5, 2000, 231–237
101. S.D.D.V. Rughooputh, H.C.S. Rughooputh: Neural network based chemical structure indexing. J. Chem. Inf. Comput. Sci. **41**, 713–717 (2001)
102. I.A. Rybak, N.A. Shevtsova, L.N. Podladchikova, A.V. Golovan: A visual cortex domain model and its use for visual information processing. Neural Networks **4**, 3–13 (1991)
103. A. Sandage, K.C. Freeman, N.R. Stokes: The intrinsic flattening of e, so, and spiral galaxies as related to galaxy formation and evolution. Astrophysical Journal **160**, 831 (1970)
104. L. Searle, R. Zinn: Composition of halo clusters and the formation of the galactic halo. Astrophysical Journal **225**, 357 (1978)
105. J.M. Siskind, Q. Morris: A maximum-likelihood approach to visual event classification. Proc. 4th European Conf. on Computer Vision, 347–360 (1996)
106. R. Somanah, S.D.D.V. Rughooputh, H.C.S. Rughooputh: Identification and classification of galaxies using a biologically-inspired neural network. Astrophys and Space Sci. **282**, 161–169 (2002)
107. R. Srinivasan, J. Kinser, M. Schamschula, J. Shamir, H.J. Caulfield: Optical syntactic pattern recognition using fuzzy scoring. Optics Letters **21** (11), 815–817 (1996)
108. R. Srinivasan, J. Kinser: A foveating-fuzzy scoring target recognition system. Pattern Recognition **31** (8), 1149–1158 (1998)
109. F. Stein, G. Medioni: Structural indexing: efficient 2-D object recognition. IEEE Trans. Patt. Anal. Mach. Intell. **14** (12), 1198–1204 (1992)
110. P. Suetens, P. Fua, A.J. Hanson: Computational strategies for object recognition. ACM Computing Surveys **24** (1), 5–61 (1992)

111. M.J. Swain, D. Ballard: Indexing via color histograms. IEEE Proc. 3rd Conf. Computer Vision, IEEE, 1390–393 (1990)
112. M.J. Swain, D. Ballard: Color indexing. Int. J. Computer Vision **7** (1), 111–32 (1991)
113. T. Tomikawa: A study of road crack detection by the meta-generic algorithm. Proc. of IEEE African '99 Int. Conf., Cape Town, 543–548 (1999)
114. S. Tominaga: A color classification method for color images using a uniform color space. IEEE CVPR, 803–807 (1990)
115. S. van de Bergh: Luminosity classification of galaxies in the revised Shapley-Ames catalog. Publications of the Astronomical Society **94**, 745 (1982)
116. A.D. Wilson, A.F. Bobick: Recognition and interpretation of parametric gesture. Proc. 6th Int. Conf. on Computer Vision, 329-336 (1998)
117. M.H. Yang, N. Ahuja: Extraction and classification of visual motion patterns for hand gesture recognition. Proc. of IEEE CVPR, Santa Barbara, 892–897 (1998)
118. A.L. Yarbus: *The Role of Eye Movements in Vision Process* (Moscow, USSR, Nauka 1965); *Eye Movements and Vision* (Plenum, NY 1968)
119. R. Forchheimer, P. Ingelhag, C. Jansson: MAPP2200 – A second generation smart optical sensor. Proc. SPIE **1659**, 2–11 (1992)
120. R. Forchheimer: Smart (optical) sensor hardware realisations. Mini-Workshop on Neural Networks for Imaging Sensors, Swedish Defence Labs, Linköping, August 1996, unpublished
121. C. Guest: University of California at San Diego. Work presented at the PCNN International Workshop, MICOM, Huntsville, AL, April 1965
122. *LAPP1110 ISA System Users Documentation*. Integrated Vision Products AB, S-583 30 (Linköping, Sweden 1997)
123. J.M. Kinser, Th. Lindblad: Implementation of the Pulse-Coupled Neural Network in a CNAPS environment. IEEE Trans. on Neural Nets **10** (3), 591–599 (1999)
124. J. Waldemark, T. Lindblad, C.S. Lindsey, K.E. Waldemark, J. Oberg, M. Millberg. Proc. SPIE **3390**, 392–402 (1998), Int. Conf. of Applications and Science of Computational Intelligence, Orlando, FL, USA, April 1998

Index

acousto-optical tunable sensor 75
aircraft recognition 43
American Sign Language 134
analogue time simulation 23
anchor velocity 104
AOTF 75
ATR 44
aurora borealis 44
Australian Sign Language 134
autocorrelation 86
Automatic Target Recognition 44
autowave 14
autowaves
   centripetal 32

Back Propagation network 44
barcodes 116
BBC 119, 120
Binary stack method 86

CCD 152
centripetal autowaves 32
ChemExper 121
ChemFinder 121
Chemical Abstracts Services 121
Chemical Directory 121
chemical indexing 121
chemical structure viewers 123
ChemIDplus 121
CNAPs 144
co-occurrence 86
composite filter 46
correlation
   binary 47
curvature flow 31

data sequences 119
de-synchronisation 14

diffraction 75
discrimination 2, 46
dynamic object isolation 58

εPCNN 69
Eckhorn model 8
edge extraction 39, 47, 52
edge frequency 86

Feedback PCNN 53, 55
feedback pulse image generator 52
Feeding 12
Field Programmable Gate Arrays 146
Fitzhugh–Nagumo model 7
Fourier filter 35
foveating 4
foveation 107
FPCNN 53
FPF 46, 55
FPGA 146
fractional power filter 46
frequencies
   higher 36
   lower 36
fusion 69, 71, 80

galaxy classification 126
Gaussian connections 18
GBC 119
generalisation 2, 46
goblet 88

hand gesture recognition 134
handwritten characters 109
HARRIS 137
Hazardous Substances Databank
   Structures 121
Hill order 123
histogram 113

Hodgkin–Huxley model 6

ICM 5, 24
image 2
   classification 2, 44
   colour 73, 95, 113
   features 35
   filtering 2
   multi-spectral 69, 75
   noisy 62
   processing 2
   recognition 2, 43, 52
   segmentation 41, 52
   shape 95
   texture 83
image factorisation 51
image fusion 69, 71
image signatures 93
interference 27, 94
iso-Dv 104

k-nearest neighbors 86
Kaiser window 77

land mines 77
LAPP 146
Law's mask 86
Learjet 44
Linking 12
   fast 16, 21
   quantized 15
Logicon Projection Network 44
LPN 44

mammography 42
MAPP2200 146
maze 115
MIG-29 44
Moody–Darken algorithm 44
motion estimation 103
multi-spectral 69, 75
multi-spectral PCNN 69

navigational systems 131
neural network 3
   feedforward 3
NIST databases 121
nucleus 88

object isolation 55
object recognition 35

optical PCNN 151
optimal viewing angle 100

Parodi model 10
PCE 56
PCNN 5
   εPCNN 69
   FPCNN 53, 55
peak to correlation energy 56
pulse capture 14
pulse image VII

Radial Basis Function 44
RBF 44
recursive image generator 55
red blood cell 41
RGB 95
road signs 131
road surface inspection 137
Run Length 86
Rybak model 9

SAAB JAS 39 43
secretion 88
shadows 60
sign language 134
signature database 99
SIMD 144
Single Instruction Multiple Data 144
SLM 152
smoothing 22
Spatial Light Modulator 152
Swiss Alps 43

TDNN 137
texture 52
Texture Operators 86
Texture Spectrum 86
thick maze 116
time signatures 16
time-delay neural network 137

van der Pol oscillator 7
VHDL 147
viewing angle 100
visual cortex 2, 4, 5
VLSI 146

wavelet 3, 80

XOR 3